Technology, International Economics, and Public Policy

AAAS Selected Symposia Series

 Published by Westview Press, Inc.
5500 Central Avenue, Boulder, Colorado

for the

 American Association for the Advancement of Science
1776 Massachusetts Avenue, N.W., Washington, D.C.

Technology, International Economics, and Public Policy

Edited by
Hugh H. Miller and Rolf R. Piekarz

AAAS Selected Symposium **68**

AAAS Selected Symposia Series

This book is based on a symposium which was held at the 1980 AAAS National
Annual Meeting in San Francisco, California, January 3-8. The symposium was
sponsored by the Society for the History of Technology.

Published in 1982 in the United States of America by
     Westview Press, Inc.
     5500 Central Avenue
     Boulder, Colorado 80301
     Frederick A. Praeger, Publisher

Library of Congress Cataloging in Publication Data
Main entry under title:
Technology, international economics, and public policy.
     (AAAS selected symposium ; 68)
     Papers presented at a symposium sponsored by the Society for the History
of Technology, held during the 1980 AAAS National Annual Meeting in San
Francisco.
     Includes bibliographical references.
     1. Technological innovations--Congresses. 2. International economic
relations--Congresses. 3. Economic policy--Congresses. 4. Science and state--
Congresses. I. Miller, Hugh H. II. Piekarz, Rolf. III. American Association
for the Advancement of Science. IV. Society for the History of Technology.
V. Series.
HC79.T4T45   338'.06   81-19803
ISBN 0-86531-319-9   AACR2

Printed and bound in the United States of America

About the Book

Rising unemployment, the reduced growth rate of productivity, a drop in international trade position--all these are symptoms of a deterioration in economic performance that currently is plaguing many industrialized nations. What impact have government policies designed to stimulate technological innovation had on this economic decline? What more can governments do to improve the situation?

This book reviews the results of five studies that have examined these questions, both in the United States and abroad. Each review describes the goals, problems, questions, and methodologies of a particular technology-policy study, critically evaluates the study's major findings and conclusions, and provides information useful to current and future deliberations about science and technology policy in the United States and other industrialized countries.

About the Series

The *AAAS Selected Symposia Series* was begun in 1977 to provide a means for more permanently recording and more widely disseminating some of the valuable material which is discussed at the AAAS Annual National Meetings. The volumes in this *Series* are based on symposia held at the Meetings which address topics of current and continuing significance, both within and among the sciences, and in the areas in which science and technology impact on public policy. The *Series* format is designed to provide for rapid dissemination of information, so the papers are not typeset but are reproduced directly from the camera-copy submitted by the authors. The papers are organized and edited by the symposium arrangers who then become the editors of the various volumes. Most papers published in this *Series* are original contributions which have not been previously published, although in some cases additional papers from other sources have been added by an editor to provide a more comprehensive view of a particular topic. Symposia may be reports of new research or reviews of established work, particularly work of an interdisciplinary nature, since the AAAS Annual Meetings typically embrace the full range of the sciences and their societal implications.

WILLIAM D. CAREY
Executive Officer
American Association for
the Advancement of Science

Contents

About the Editors and Authors.....................xi

Preface..xiv

1 U.S. Technological Innovation and the
   Nation's Competitiveness in International
   Trade--*Edward M. Graham*.........................1

   Reasons for U.S. Comparative
      Advantage in Post-World
      War II World Trade for
      Technology-Intensive Prod-
      ucts and Reasons Why This
      Advantage May Be Dimin-
      ishing   2
   U.S. Dominance of Technological
   Innovation Following World War
   II,2; The Decline in U.S. Domin-
   ance,7; Is There Evidence Indi-
   cating a Loss of U.S. Comparative
   Advantage Based upon Technological
   Innovation?,19
   Choices for the United States   22
   References   27
   Bibliography   28

2 Industrial Innovation and Government
   Policy--*Richard A. Meserve*.....................31

   Appendix A: President Carter's
      Message to Congress, October
      31, 1979   39
   Appendix B: The President's
      Industrial Innovation Initia-
      tives, October 31, 1979   45

3   Technical Capability and Industrial
     Competence --*Gunnar Hambraeus*......................61

     Introduction   61
     The Historical Development of
        Swedish Industry   63
     The Status of Swedish Education
        and Research   65
     Swedish Industry Today   66
     Trends in International Trade
        and Industry   67
     New Competitors and Markets for
        Sweden   68
     World Trends in Technological
        Development   70
     Conclusions and Proposals   71
        Sweden's Adaptation to Change-
        improving Industrial Performance,
        71; Strengthening the Knowledge
        Base,73; New Technological
        Prospects,75; Governmental Support
        and Control Measures,75; Long-term
        Studies and Research and Development
        Planning,76

4   The Committee for Economic Development
     Report on United States Technology Policy --
     Robert C. Holland 79

     Background of the CED Study of
        Technology   79
     The Contribution of Greater
        Technological Innovation   81
     Identifying the Innovation
        Problem   84
     Policy Strategy to Stimulate
        Technological Progress   87
     Followup to the CED Study   93
     Conclusion   95
     Appendix A.   Subcommittee on
        Technology Policy in the
        United States   97
     References   100

5   Research and Innovation--*Walter A. Hahn,*
     Mary Ellen Mogee 101

     Introduction and Background   101
     Industrial Innovation as a
        Public Policy Issue   102

The RIAS Final Report   106
  What Is Known?,107; Status,109;
  Explorations,112; Outlook,114
Selected Findings   117
Implications for the Congress   123

6   The International Dimensions of Innovation--
     David Z. Beckler 127

     Structural Factors in Innova-
       tion   130
       Regulatory Policies,130; Protec-
       tionism,131; Technology Transfer
       Policies,131; Manpower and Social
       Policies,131; Patent Policy,132;
       Small- and Medium-sized Firms,132;
       Innovation in the Service Sector,
       132; Diffusion of Innovation,133;
       Public Acceptance,133; Direction
       of Technological Change,133; Scien-
       tific and Technical Inputs,134
     Selective Intervention   134
       Enterprise Level,134; Sectoral Level,
       135; "Picking the Winners",135
     Scientific and Technological
       Inputs to Innovation   136
     Government Policy and Structures
       for Innovation   137
       Technology Policy,137; Integration
       of S&T and Economic and Social
       Policies,137
     International Cooperation in
       Promoting Innovation   138
       Role of the OECD, 138

Discussion: U.S. Technological Policy Needs:
Some Basic Misconceptions--*Bela Gold* 141

Discussion--*N. Bruce Hannay*........................145

Discussion--*Nathan Rosenberg*.......................149

About the Editors and Authors

Hugh H. Miller *is executive director of the Office of the Foreign Secretary, National Academy of Engineering, where he has been responsible for studies on technology, socioeconomic development, and international competition; the impact of U.S. policies on industrial innovation and competitiveness; and the international transfer of technology.*

Rolf R. Piekarz, *a specialist in international economics, is a senior staff associate, Division of Policy Research Analysis, National Science Foundation, where he heads a group which is concerned with how federal policies modify the rate and direction of research, development, and technological innovation; what the economic and social consequences of these modifications are; and what the economic consequences are of possible U.S. responses to changing conditions in international trade and finance. He is the editor of* International Policy Research *and* The Effects of International Technology Transfers on the U.S. Economy *(National Science Foundation, 1981 and 1974, respectively).*

David Z. Beckler, *now director for Science, Technology and Industry for the Organization for Economic Cooperation and Development, was formerly executive officer for the President's Science Advisory Committee and assistant to the Science Advisors to the President.*

Bela Gold *is director of the Research Program in Industrial Economics, chairman of the College on the Management of Technological Change of the Institute of Management Sciences, and William E. Umstattd Professor of Industrial Economics at Case Western Reserve University. He is also president of Industrial Economics and Management Associates, Inc. He serves on numerous committees devoted to the study of domestic international economic problems and as an industrial consultant. Among his numerous publications are* Appraising and Stimulating

Technological Advances in Industry *(Oxford: Pergamon, 1980) and* Productivity, Technology and Capital: Economic Analysis, Managerial Strategies and Governmental Policies *(Lexington, 1979).*

Edward M. Graham, *an assistant professor in the Sloan School of Management, Massachusetts Institute of Technology at the time this paper was written, is now on the staff of the Organization for Economic Cooperation and Development. A specialist in international economics, he is principal author of a National Research Committee report,* Technology, Trade and the U.S. Economy *(National Academy of Sciences, 1978).*

Walter A. Hahn *is a senior specialist in science, technology and futures research at the Congressional Research Service. He has written on science policy and innovation and is coeditor of* Assessing the Future and Policy Planning *(with K.F. Gorden; Gorden and Breach, 1973).*

Gunnar Hambraeus, *currently professor and president of the Royal Swedish Academy of Engineering Sciences, has long been active in the field of scientific research and technology policy. He served as chairman of the IIASA Advisory Committee for the management of technology, and was secretary of the Swedish Technical Research Council, editor and publisher of the journal* Teknisk Tidskrift, *managing director of the Swedish Technical Press, and consultant to the International Atomic Energy Agency.*

N. Bruce Hannay, *a physical chemist by training, is vice president for research at Bell Laboratories and Foreign Secretary for the National Academy of Engineering. The author of numerous technical articles and books, he has served as president of the Industrial Research Institute, chairman of the directors of Industrial Research, and president of the Electrochemical Society, and he is a member of the National Academy of Sciences and the American Academy of Arts and Sciences. He has served a number of universities and government agencies in an advisory capacity.*

Robert C. Holland, *a specialist in financial economics, is president of the Committee for Economic Development, Washington, D.C., which has recently published* Stimulating Technological Progress *(1980).*

Richard A. Meserve, *a physicist by training, is currently an associate in the law firm of Covington and Burling in Washington, D.C. From 1977-81 he served as senior policy analyst and legal counsel in the Office of Science and*

Technology Policy where he was concerned with issues affecting energy, industrial innovation, and the support of research and development.

Mary Ellen Mogee, *currently a technology policy analyst in the Experimental Technology Incentives Program at the National Bureau of Standards, has studied various aspects of technology policy, including innovation and regulation, and technology, trade and productivity. She is the author of* Technology and Trade: Some Indicators of the State of U.S. Industrial Innovation *(U.S. Congress Ways and Means Committee, 1980) and* Industrial Innovation and Public Policy Options: Report of a Colloquium *(National Academy of Engineering, 1980).*

Nathan Rosenberg, *a specialist in the economics of technology, is a professor of economics at Stanford University. He is the author of* Perspectives on Technology *(Cambridge University Press, 1976) and* Technology and American Economic Growth *(M.E. Sharpe, 1972) and is a former editor of the* Journal of Economic History.

Preface

This volume presents the papers and prepared remarks discussed
during the session, "Technology, International Economics, and
Public Policy," held at the 1980 AAAS Annual Meeting. The objective of the session was to review a sample of recent government sponsored and private sector studies conducted in the
United States and elsewhere about the contribution of public
policy in an industrial nation to technological innovation
for the purpose of enhancing national economic performance.
These studies usually attempt to make some consensus judgments about the role of government policy in stimulating
economic growth through technological innovation and to suggest modifications in public policies to increase technological innovation, especially in industry.

These exercises at soul searching have occurred against a
background of deterioration since the beginning of this decade in the economic performance among all industrial nations.
This deterioration has been measured in terms of a number of
indicators of economic performance. Some industrial nations
have experienced a sharp rise and secular increase in unemployment over the course of the decade. Other industrial
nations have witnessed a decline in their international trade
position. Most of the developed nations have been exposed to
a decline in the growth of their economies and a reduction in
productivity growth. In discussions about deteriorating
economic performance, the role of technological development
and technological change play a prominent role. Perception
of diminished economic standing and performance and the role
of lower technological innovation in the poorer economic performance have been especially keen in the United States. The
paper, "U.S. Technological Innovation and the Nation's Competitiveness in International Trade," by Edward M. Graham,
uses available research and data to make some balanced judgments about the U.S. situation.

Deliberations about the role of technological innovation in economic performance usually presume that government policies play an important role in the technological development of industrial societies. A number of examples are frequently cited. Government provides some form of tax incentives for research and development and for investments in new plants and equipment. All nations have legal systems which give special protection to intellectual property rights. Governments underwrite research at universities and provide subsidies to encourage firms and other institutions for applied research and development. Therefore, if technological innovations for economic purposes appear to lag, one of the most likely places to look for improvement is in government policies. Such an approach introduces two obvious questions: What has been wrong with government policy? What can governments do to improve the situation?

These concerns and questions set in motion during the past three to four years a number of studies in the United States and abroad to look closely at what governments are and should be doing to stimulate technological innovation, especially in industry. Here in the United States, the Federal government sponsored two studies - one conducted by the Commerce Department under the supervision of the Office of Science and Technology Policy in the Executive Office of the President; the other conducted by the Congressional Research Service under the supervision of the Joint Economic Committee of the U.S. Congress. On the private sector side, the Committee for Economic Development examined U.S. policies with regard to technological innovation. In other industrial nations, the governments of Canada, Germany, Japan, Sweden, and the United Kingdom all undertook or sponsored efforts to look at the influence of the government's policies on technological innovation. On an international level, the Secretariat at the Organization for Economic Cooperation and Development (OECD) looked at the economic, social, and policy environment for technological innovation among industrial nations.

Summaries and selected reviews of a sample of these studies may provide information useful to current and future deliberations about science and technology policy. Given continuing interest about the role and influence of government policies on technological innovation, these summaries present an historical record of perceptions among different groups in various localities at a particular time about shortcomings and about improvements in government policies. Looking at a number of studies also yields some insights about strengths and limitations of different approaches for examining policies. Finally, reports about some of the policy exercises will highlight differences and similarities in views about problems relating to technological innovation for enhancing economic growth and international trade.

In preparing the papers about the policy exercises, the authors were asked to use the following common format. First, the paper was to set forth the author's perspective of the study goals and the problems and questions with which the study was intended to deal. Second, the author was to describe how the study was organized and what methods were used to obtain answers and to develop the policy recommendations. Third, the paper was to discuss the major findings and conclusions of the study.

In addition to the background paper by E. M. Graham* from the U.S. Department of the Treasury, summaries of five policy studies have been prepared. These summaries cover all three U.S. exercises on innovation policy and the policy studies prepared by the Royal Swedish Academy of Engineering Sciences and the Secretariat of the Organization for Economic Cooperation and Development. Richard Meserve, from the Executive Office of the President's Office of Science and Technology Policy, presents the sketch of the policy review conducted by the Commerce Department in the paper, "U.S. Efforts to Utilize Science and Technology for Industrial Innovation." The U.S. government policy study conducted by the Congressional Research Service of the Library of Congress is described by Mary Ellen Mogee and Walter A. Hahn in "Report on the Research and Innovation Area Study: Joint Economic Committee Special Study on Economic Change." Robert C. Holland, President, Committee for Economic Development, reports on the study by the Committee for Economic Development task force, which he headed, in the paper, "Report on U.S. Technology." The Swedish policy exercise is described by Gunnar A. Hambraeus, President of the Royal Swedish Academy of Engineering Sciences, in the paper, "Technical Capability and Industrial Competence." Finally, the OECD effort is reported in "Description of the OECD Studies and Results on Technological Innovations in Large, Small, and Medium-size Firms," by David Z. Beckler, Director for Science, Technology and Industry at the OECD.

Hugh H. Miller
National Academy of Engineering

Rolf R. Piekarz
National Science Foundation

---

*Now with the Organization for Economic Cooperation and Development, Paris.

Edward M. Graham

1. U.S. Technological Innovation and the Nation's Competitiveness in International Trade

International trade theory teaches that the basis for trade between nation-states lies in the principle of comparative advantage. This principle asserts that a nation should export goods which it can produce at a cost relatively lower than can other nations, and, conversely, the nation should import those goods which would cost relatively more to produce domestically. The notion of relativity is important to the principle. Even if a nation could, in absolute real resource terms, produce all goods more cheaply than a second nation, it might still benefit the first nation to import from the second those goods which are relatively costlier to produce at home. The case for free trade among nations rests largely on the demonstration that world wealth is maximized if each nation specializes in the production of goods in which it possesses a relative cost advantage and trades with other nations to fulfill some or all of domestic demand for other goods.

The reasons why a nation might possess a comparative advantage (i.e., relative cost advantage) for certain goods are several. Classical economic theory stresses endowments of factors of production, demonstrating that under assumptions of static technology and equivalent production functions among nations, a nation will possess comparative advantage in goods requiring intensive use of those factors in which the nation is relatively well endowed. Thus, a nation characterized by a large quantity of arable land per head of population might be expected to have a comparative advantage in agricultural products, while a nation having little land per head of population might possess a comparative advantage in labor intensive manufacturing. If technological capabilities of nations are unequal, however, one would expect nations with greater capabilities to have a comparative advantage in those goods which embody advanced technologies. For reasons to be discussed later in this paper, an advantage accruing to a specific technology may be quite short-lived, and to sustain an overall

1

comparative advantage in "technology-intensive" goods, the nation would have to renew continually its advantage by innovating new technologies.

During the 1960's and early 1970's, researchers amassed a vast amount of data pointing to technological innovation as a significant determinant of U.S. competitiveness in the export of manufactured products, thus "revealing" a comparative advantage of the nation in technology-intensive goods (1). An issue to be addressed in this paper is whether the nation might be losing this advantage, and, if so, of what concern is the loss.

Prior to addressing this issue, it is of use to examine, first, why the United States has held such an advantage in the post-World War II era and, second, why one might expect the relative advantage of the nation to be in decline at the present time.

I. Reasons for U.S. Comparative Advantage in Post-World War II World Trade for Technology-Intensive Products and Reasons Why This Advantage May Be Diminishing

   A. U.S. Dominance of Technological Innovation Following World War II

Possibly the best explanation of U.S. comparative advantage in world trade of technology-intensive goods is embodied in what has become known as the "product life cycle model." This model was first outlined in comprehensive detail by Raymond Vernon during the 1960's, although a number of individuals had earlier published key ideas which were to be incorporated into its formulation (2). The model has been tested and refined by a number of persons, including Louis Wells, Robert Stobaugh, and John E. Tilton (3).

Under Vernon's formulation of the model, internal market characteristics unique to the United States are deemed to be the causal element behind U.S. advantage in product innovation. Three characteristics can be identified.

First is the very high per capita income and the large number of high income consumers of the United States. These characteristics, it is argued, induce U.S. entrepreneurs to develop new types of consumer products for which demand is highly income elastic. These products, generally classifiable as "luxury" goods, include the transistor radio of the 1950's, the electronic hand-held calculator of the late 1960's, and the electronic watch of the middle 1970's. In each case, it should be noted, the good ceased to be a "luxury" within a

few years of its introduction, a point which will be taken up
shortly.

Second, the relative cost of labor to capital in the
United States has been high compared to other nations. Vernon
reasoned that this would induce U.S. firms to develop labor-
saving capital goods. The possibility that factor cost relationships would induce factor biased innovation has long
been discussed by economists, who have theorized that in nations where the relative cost of specific factors of production is high, technological innovation will be biased toward
use of the cheaper factor (4). Significantly, empirical
studies conclusively demonstrating that technological innovation in the United States is biased toward the use of capital
(i.e., labor saving) have not been published, although one
by William Davidson is clearly suggestive of this possibility
(5). Vernon simply assumed that high labor costs in the
United States would induce the development of labor-saving
goods.

Third, the United States is a very large market, both in
terms of aggregate consumption and geographic extent. This,
Vernon argued, would induce U.S. firms to develop new products
which would be characterized by economies of scale. Two categories of products can be identified. The first encompasses
products which can be most economically manufactured on a
large scale, for example, many petrochemicals. The second
category subsumes capital goods products that themselves enable the achievement of scale economies in the manufacture of
other goods. In addition, innovation of goods such as transport and telecommunications equipment that enable producers
to link two geographically separate markets and service them
from single manufacturing or distribution centers might be
induced.

These three general characteristics of the U.S. market
are unique in the sense that the nation has, throughout most
of this century, been richer and better endowed with capital
on a per capita basis and has possessed a larger aggregate
internal market than any other nation on the earth. This was
especially so during the first twenty years or so following
the conclusion of the Second World War, during which time most
other large industrial nations faced the task of rebuilding
capital stocks and revitalizing their economies. During this
time, U.S. firms could innovate new products in response to
market conditions at home and could also expect to find growing markets for these same products in the recovering economies
of Western Europe.

A firm in possession of a unique new product is often

in the position of the proverbial inventor of the "better mousetrap": the world beats a path to the inventor's door. In time, however, this situation changes. New competitors enter the market, and if the inventor is not careful, his favored position can be undermined.

In part, changes in the competitive situation result from changes in the product design itself, resulting from information feedback from the marketplace which enables designers to determine exactly the needs of the consumer. As these needs become better understood, the product can be both improved and standardized. Standardization of the product allows changes to be made in its manufacture. While manufacture of a nonstandardized product requires a process which utilizes adaptable production equipment and skilled labor as the design becomes standardized, specialized production equipment can be substituted for the flexible equipment and less skilled for skilled labor. Highly specialized, often automated production processes typically require less labor input per unit of output than do less specialized processes, and hence, a net substitution of capital for labor for a given level of output is achieved. One result of these changes is (usually) to reduce the marginal cost of producing the product (6).

Changes in the competitive structure of the supplying industry come about as competing firms enter the market for the newly innovated good. Once a new product has been successfully launched in the marketplace, the innovator will typically earn a monopoly rent on sales. Indeed, it is argued that the prospect of the rent is what induces the innovation in the first place (7). Seeking to capture some of the rent, other firms will enter the market with competing products. Ease of entry is enhanced as the product becomes standardized. Because the innovator firm will have overcome many of the uncertainties associated with the introduction and development of a new product, imitator firms will not have to bear the costs associated with these uncertainties. Thus, a "demonstration effect" of successful innovation can serve as an incentive for imitating firms to enter the market. To be sure, one counteracting disincentive might be patents held by the innovator. In most cases, however, the patents can be "invented around" by the determined imitator. The competition engendered by new firm entry will drive down the price of the product, although the rent to the innovator firm may or may not be reduced depending upon whether it can lower the cost of manufacture of the increasingly standardized product faster than the competitors. Thus, innovation and subsequent new entry stimulate cost reduction of the new product.

Demand for the product will change as information about it becomes diffused. Initial demand for a new product may be constrained by uncertainty (or just plain ignorance) about the uses to which the product can be put or its performance. As the product becomes more widely used, however, additional potential users of the product will learn of its uses and will consider its purchase. Additionally, if demand is price elastic, price reductions will lead to growing sales revenues. In the longer run, the structure of demand might be affected by rising per capita incomes. Louis Wells has suggested, for example, that demand for U.S. innovations in Europe has lagged U.S. demand during the 1950's and 1960's because incomes in Europe have lagged those in the United States (8).

Supply and demand changes will come about in overseas markets as well as domestic ones. As information about the product becomes more widespread, overseas demand for the product will grow. If overseas demand is price elastic, a decline in the price of the product will result in further growth in volume of overseas sales. Local firms in foreign nation markets will be induced to imitate the product and commence local production. Vernon has postulated that in order to avoid loss of overseas marketshare to local producers, U.S. exporters might themselves become "local" producers by establishing manufacturing subsidiaries within the foreign markets. Therefore, he argued, U.S. manufacturing firms become multinational for defensive reasons (9).

Based on the above reasoning, Vernon and Wells developed a temporal sequence of stages of the "product life cycle" in the export of U.S. new product innovations. Early on, the product would be manufactured and consumed solely in the United States. In time, however, demand would appear in other advanced economies such as those of Europe and Japan. Overseas demand would at first be filled by U.S. exports, but eventually these exports would be displaced by local production in other advanced nations. In some cases, the "local" producers would be subsidiaries of U.S. firms. Demand eventually would begin to appear in developing nations and would be met initially by either European or U.S. export. Eventually, exports from other advanced nations would capture a share of the U.S. market, and ultimately the locus of production might shift to developing nations. At some point, these nations might export to both the United States and to other advanced nations.

As will be noted in the following section of this paper, the notion that the United States is the sole center of new product innovation as depicted in this temporal sequence is certainly now obsolete (10). However, the described sequence

can provide insight into the diffusion of new product technology from any innovating nation.

While Vernon and his colleagues look primarily to the
characteristics of demand in the United States to explain the
country's historically high rate of innovation, another
analyst, Burton Klein, focuses on the characteristics of supply
(11). Klein hypothesizes that the firm, when confronted with
the decision of whether to invest resources in innovation,
faces two types of uncertainty. "Type I uncertainty" is the
uncertainty intrinsic to innovation. A would-be innovator is
uncertain of levels and elasticities of demand for a new product. The innovator is also uncertain of the costs involved
in perfecting the new product. If the product represents a
major departure from existing products, the innovator might
even be uncertain of the design and performance characteristics of the product. These uncertainties pose risks to the
firm since there is some possibility that the investment in
innovation will yield unsatisfactory or negative returns.
Thus, the uncertainties act as disincentives to innovation.
Offsetting these "Type I uncertainties" is the potential of
reward should the innovation prove successful.

"Type II uncertainty" is uncertainty with respect to the
actions of competitors. If a competitor firm is first to
develop a new technology, that firm will possess an advantage
in the marketplace over its rivals. The possibility that a
competitor firm will preempt a new technology acts as an incentive for other firms to engage in innovation.

Klein argues that both the potential of economic reward
for overcoming "Type I uncertainty" and a high level of "Type
II uncertainty" are prerequisites for innovation to occur at
a rapid pace in a market economy. "Type II uncertainty" is
high if rivalry within an industry is intense. Rivalry exists
when competing firms are unable effectively to collude with
one another on either an overt or a tacit basis. Klein believes that the most important prerequisite for a high degree
of intraindustrial rivalry is easy entry into the industry.
Generally, entry into an industry is facilitated if (1) the
industry is rapidly growing; (2) scale economies are not of
crucial importance; (3) key technologies are not closely held
by one or a few firms; and (4) other important intangible
assets (such as brand names) are not held by one or a few
firms.

Klein notes that highly innovative firms are characterized by internal organizations that are not rigidly hierarchi-
cal. Innovation is most likely to be forthcoming from an
organization in which managerial and technical employees are

able to interact freely with one another on a personal basis.
He observes that firms operating in highly rivalistic industries typically possess interactive internal organizations.

It should be noted that Klein sees his hypotheses not so
much as an alternative but as a supplement to the product life
cycle. The Vernon version of the product life cycle, in
Klein's view, places too much emphasis on the demand side of
the U.S. economy in its explanation of U.S. innovation, while
the characteristics of the supply are an equal, if not more
important, determinant of innovation. In particular, he
points out that Vernon's arguments do not explain why some
sectors of the U.S. economy are technologically dynamic while
other sectors are virtually moribund.

B. The Decline in U.S. Dominance

The product life cycle model, as supplemented by Klein,
provides a basis for explaining the dominance of the United
States in generating new commercial technologies during the
aftermath of World War II. As has been noted, the era was one
when per capita incomes in the United States were much greater
than those of other large industrial nations and when rivalry
in high technology industries was largely limited to U.S. firms.
Clearly, however, this era has ended, and characteristics once
unique to the U.S. market are now evident in other nations as
well. In this section, it is argued that at least four major
trends are significantly affecting U.S. comparative advantage
in world trade of goods embodying new technologies. These can
be summarized as follows:

First, the technological capabilities of a number of
industrialized nations other than the United States have advanced very rapidly during the past fifteen years or so. In
certain sectors these nations, and not the United States, have
become the leading sources of new technologies. Second, the
industrial sectors of a limited number of the so-called developing nations have grown very rapidly in recent years.
Several of these nations are beginning to emerge as important
exporters of certain types of manufactured products. Third,
the ability of multinational firms to transfer new product
technologies across international lines has been growing.
Additionally, the ability of local firms in a number of foreign
nations to quickly imitate new technologies has increased
markedly. Fourth, in the last ten years or so, there is some
opinion holding that a worldwide slowdown in the rate of development of new industrial technology has occurred, although
there is no unequivocable evidence of this.

In the following paragraphs, the causes and consequences

Table 1. Total Expenditures on Research and Development of Five Major Nations, 1963-1977 (in billions of dollars).

	1963	1967	1969	1971	1973	1975	1977
France	1.3	2.5	2.6	3.2	4.2	5.8	7.1
W. Germany	1.4	2.4	3.3	5.5	7.6	9.4	12.9
Japan	1.0	1.4	2.9	4.9	7.9	9.8	NA
United Kingdom	2.1(est.)	2.3	2.5	2.9(est.)	NA	4.3	NA
United States	13.4	23.6	26.2	27.3	30.6	35.2	47.3
TOTAL	23.2(est.)	32.2	37.5	43.8	NA	64.5	NA
U.S. as a % of total in 1975	75	73	70	62	NA	54	NA

Source: National Science Foundation, Science Indicators, 1978 (U.S. GPO, 1979), Table 1-1; figures converted to $US at exchange rate prevailing at year end; estimates for United Kingdom for 1963 and 1969 supplied by Embassy of the United Kingdom to the United States.

of each of these trends are explored. It is important to note in this discussion that the effects of non-technological developments in the world economy (such as the oil price increase of 1973, to choose the most apparent example) are difficult to separate from the effects of technology-related developments. It is beyond the scope of this discussion to attempt to explore all such effects. The comparatively modest goal here is to identify a limited number of the most important factors affecting world trade in technology-intense goods.

- **Growing Technological Capabilities of Other Advanced Nations.** That the technological capabilities of a certain number of industrialized nations other than the United States have advanced rapidly in recent years should not, at base, be considered a surprising development. Nations such as the United Kingdom, Germany, and France have, of course, historically been major sources of technological innovation. However, the economies of these nations were dealt more severe blows by World War II than was the U.S. economy, and as a result the United States emerged as the dominant source of new technological innovation even in sectors which had been dominated by other nations prior to the war. Even by the middle 1950's, however, it was evident that the economies of these nations would recover and that their previous eminence in at least some sectors would be reattained.

  The recovery from the war was fairly complete in Western Europe and Japan by the late 1960's. Table 1 indicates that by this time a growing percentage of the world's research and development activity was being performed outside the United States. To be sure, research and development is but one input into the process of technological innovation, and the table does not indicate innovative output. Nonetheless, to the extent that input can be taken to be a proxy for output, the table shows that the relative importance of the United States as an innovator declined steadily from the 1950's through the late 1970's.

  Of the war-devastated economies, the two that emerged as the strongest were those of the principal foes of the United States during World War II, West Germany and Japan.

  The redeveloped economy of West Germany is in many regards similar to that of the old Germany. Sectors in which Germany had been historically strong reemerged as those in which the West German state has become a leading innovator. These include the chemical and pharmaceutical industries, precision and heavy machinery, heavy electrical goods, metallurgy, and surface transport equipment. West German firms generally have not yet emerged as leading innovators in very high technology industries like aircraft and aerospace, advanced electronics, and high speed electronic computation. Some German firms have achieved excellence in certain high technology subsectors, however, especially in chemicals, pharmaceuticals, nuclear power, and telecommunications.

Japan has perhaps been the major technological success story of the latter half of the 20th century. In the early postwar period Japan rebuilt her traditional industries, becoming a major exporter of cotton textiles and simple consumer products. Rebuilding of the steel, shipbuilding, and heavy machinery industries followed, and in these sectors Japan became a very efficient producer (and exporter) during the early to middle 1960's. The automotive and consumer electronics industries were built up rapidly during the 1960's and early 1970's. Japanese firms meanwhile were making advances into such high technology industries as semiconductor and computer manufacture. The Japanese Ministry of International Trade and Industry (MITI) now expects that a major source of future growth in the domestic economy will come from the most technologically advanced sectors. This has led to a major effort to surpass the United States in the development of ultra-sophisticated microelectronic technologies and high speed computers.

Implications for the United States of advances in other industrial nations' technological capabilities are straightforward. The percentage of the world's commercial technological innovation originating in the United States began to decline during the early 1960's, and this decline has continued into the 1970's. In some sectors in which the United States dominated the introduction of new technologies in the early post-World War II years, the dominance has ended, and in certain sectors, U.S. industry has ceased to play much of an innovative role at all. Monopoly rents that U.S. firms could once achieve from export sales of technologically advanced products have in many cases been reduced substantially or disappeared entirely.

One consequence has been that the terms of trade in

manufactured goods have turned against the United
States. Dollar exchange rates, established under the
Bretton Woods Agreement immediately following World
War II with the express purpose of encouraging U.S.
investment in war-torn areas, also enabled the United
States to import many manufactured goods cheaply for
several decades. As the uniqueness of U.S. exports
diminished, a shift in the terms of trade put pressure
on the parity of the dollar. This pressure might have
contributed in part to the collapse of the Bretton
Woods system during the early 1970's.

One important aspect of the technological resurgence
of Western Europe and Japan requires special mention.
It has already been noted that factor-saving technological innovation is hypothesized to act so as to
conserve on scarce factors. It has been noted further
that historically the scarcest factor in the United
States has been skilled labor, and thus that U.S.
factor-saving innovation has been labor-saving. In
both Western Europe and Japan, energy has historically
been a scarce factor, and consequently much innovation
in these areas has been directed toward the conserva-
tion of energy. This is probably most obvious in the
case of motor vehicles, but is also true of space-heat-
ing technologies and industrial process technologies.
Energy has, of course, recently also become scarce in
the United States, and therefore energy-saving technologies already developed outside North America are
increasingly relevant to the needs of North America.
It is both likely and desirable that these needs will
cause imports into the United States of foreign-made
goods embodying energy-saving technologies to increase
as well as imitation of foreign technologies by U.S.
producers to accelerate. Just as rising per capita
incomes in other advanced nations during the 1950's
and 1960's and the resulting export boom for U.S. producers of products of advanced technologies benefited
both producer and consumer nations, so should the use
of foreign technologies be seen as an opportunity and
a challenge to the nation. More is said on this matter
later in this paper.

Growing competition among the firms of industrialized
nations to supply products embodying advanced technologies will doubtlessly lead to emphasis on product
differentiation in some industries. A Swedish economist, Staffan Linder, predicted in the early 1960's
that product differentiation would become a major
determinant of international trade among affluent

economies (12). Linder's assertion was that differing
consumer tastes among nations would cause local producers in each nation to develop products tailored to
meet those tastes but which would be but slightly
differentiated from products of equivalent function
developed in other cultures. Because in every country
there would be some local demand for those variants
of a product produced in other countries, there would
arise international trade for similar but differentiated goods. For example, the United States might import Volkswagens from Germany but, at the same time,
export Cadillacs to Germany. One consequence to trade
in differentiated products demonstrated by Linder was
that the gains from such trade are much less than from
trade of dissimilar products. Thus, one determinant
of U.S. competitiveness in world trade may become the
ability of U.S. firms to develop differentiated products which appeal to foreign consumers.

Paradoxically, however, U.S. competitiveness in some
industries may rest on less product differentiation
rather than more. This is especially the case for
energy utilization technologies, where product differentiation between U.S. and foreign manufacturers has
resulted from significantly lower costs of energy to
consumers in the United States than abroad. From the
perspective of 1981, it is difficult to comprehend
that it was only about a quarter of a century ago that
the United States was a net exporter of energy. The
need to adjust to higher domestic energy costs has
forced U.S. manufacturers in a number of industries
to make significant product design changes. The cost
of adjustment has adversely affected the fortunes of
a number of firms, but there should be little doubt
that the adjustment must be made.

It must be noted, in spite of the foregoing paragraphs,
that U.S. manufacturing firms in many sectors remain
quite innovative and internationally competitive. This
is particularly true in industries characterized by
very new and rapidly changing technologies. For U.S.
firms operating in these industries, the economies of
Western Europe and Japan continue to offer significant
export opportunities, because they continue to require
the products of advanced U.S. technologies. Depreciation of the dollar against foreign currency has
almost surely increased competitiveness in world
trade of at least some U.S.-made technology-intensive goods.

● <u>Industrialization of Developing Nations</u>. Rapid industrialization in developing nations poses a different
set of implications for the United States than does
the technological progress of other advanced nations.
Rapidly industrializing nations possess labor forces
for which wages are much lower than in industrialized
nations, even after adjustment is made for differences
in skill and productivity. Only a relatively small
number of the world's 100 or so developing nations,
however, are becoming significant exporters of manufactured products. These products are generally those
characterized as embodying mature technologies. Furthermore, only a few of these nations, most importantly
Brazil, Mexico, South Korea, and India, are on the way
to developing broadly diversified industrial bases.
In other nations. industrial activities tend to be
quite specialized. The major manufactured products of
the rapidly industrializing group currently include
textiles, apparel, shoes, and standardized consumer
electronics. Small but growing product categories include steel and steel products, standardized machinery,
ships, and automotive components (13).

Industrialization of developing nations presents the
United States with a number of dilemmas and important
choices. Because of low labor costs, developing nations are becoming competitive largely in world trade
of goods based on mature or standardized technologies.
It is clear that the more labor input required per unit
of output for a good, the more competitive low wage
nations can become as producers of that good. In the
absence of governmentally imposed barriers to trade,
labor intensive goods from developing nations will inevitably capture a growing share of world markets,
including the U.S. domestic market.

Equally inevitable will be calls for border restrictions
to protect threatened industries. Such calls should
not be heeded. To protect these industries would be
to force U.S. consumers to pay high domestic prices for
domestically produced goods which could be more cheaply
purchased from abroad, tantamount to taxing U.S. consumers to subsidize producers. Furthermore, protection
of a domestic industry shields producers in that industry from the stimulus of foreign rivalry. The
shield acts as a disincentive to innovation within the
industry, eventually causing the industry to lag behind world standards in the introduction and use of
new technology. Thus, protection of a domestic

industry ultimately results in technological obsolescence within that industry.

Worse yet from a domestic economic point of view,
protection of import threatened industries would prevent a transfer of labor and capital out of these and
into more dynamic industries. The long-run effect of
this for the nation as a whole would be slower growth
and increasing backwardness. Even for workers in the
protected industries, growth of incomes in the long
run would be retarded even if in the short run jobs
were to be saved.

Politically, the effects of large-scale curtailment
of imports of manufactured goods into the United States
would be highly negative. Throughout the post-World
War II era, the United States has been an advocate of
economic development in the Third World. Most of the
rapidly industrializing Third World nations are highly
dependent upon export markets to earn foreign exchange
needed for future growth. This is especially so in
the wake of higher oil prices. For the United States
to adopt protectionist policies toward these nations
would undermine their prosperity and create new re-
sentments. The United States would be seen as a nation
willing to "beggar its neighbors" in order to serve
its own narrow interests.

With respect to the newly industrializing nations themselves, a number of changes will occur as industrialization progresses. Probably the most important change
will be in wage rates. As these nations become more
affluent, the differential between local wages and
those paid in the industrialized world will narrow.
Eventually, the comparative advantage in labor-intensive manufactured goods currently possessed by these
countries will disappear, shifting perhaps to what are
now the very poorest nations. To continue their ascen-
dancy, the newly industrializing nations themselves
will have to transform their industries, allowing
mature sectors to contract to release resources into
more dynamic sectors. This need, however, lies well
into the future.

● Transfer of Technology by Multinational Firms. Multinational firms are able to transfer technologies
  rapidly across national boundaries, and as activities
  of these firms continue to grow, trade advantages
  accruing to the United States from new product innovation are likely to become increasingly short-lived.

There is some evidence that the speed at which multinationals transfer technologies is increasing. For
example, empirical data recently collected by Raymond
Vernon and William Davidson indicate that the time
lag between the introduction of a new product in the
United States by a U.S. multinational and its first
use abroad by a foreign affiliate has steadily decreased during the past three decades (14). In addition, of course, non-U.S. firms, especially those of
Japan, have exhibited increasing capabilities to
rapidly imitate new technologies first introduced in
the United States. Thus, the length of time that the
United States can expect to export products embodying
new technologies prior to the commencement of production abroad is decreasing.

Potentially negative effects on the U.S. economy of
rapid diffusion of new product technology can be
identified. The U.S. economy bears most of the costs
of the domestic creation of these new product technologies but may yield much of the benefit to nations
to which the technologies are transferred if full com-
pensation for the transfer is not received. If the
total benefit that can be appropriated by a U.S. firm
is further reduced by the rapid diffusion, there will
be less incentive for innovation to occur. This is
likely to be the case if the diffusion is accompanied
by rapid imitation of the technology by the worldwide
competitors of the innovating firm. This is not, however, invariably the case. If the innovating firm is
able to utilize its own network of subsidiaries to
diffuse the technology and is able to exclude the technology from competitors for a certain length of time,
the diffusion might act as a stimulus to innovation.
This might be the case, for example, if the innovating
firm were able to achieve very rapid cost and price
decreases by quickly transferring the technology to
subsidiaries located in low-wage areas (15).

There have been numerous proposals to regulate the
transfer of technology out of the U.S. economy, but
such proposals have invariably run into problems of
practicality. Most efforts at controlling technology
outflow, viewed with historical perspective, have
proven to be exercises in "closing the barn door after
the horses have fled" (16). Past efforts apart,
whether diffusion of technology can practicably be
retarded is a vital question. Technology is human
knowledge and to prevent one determined human being
from acquiring the knowledge of another is not easy.

Table 2.   The 187 Largest U.S. Multinational Firms Classified by Industry.*

Industry (Standard Industrial Classification Number)	Number of Firms
Food and Kindred Products (20)	29
Tobacco Products (21)	1
Textile Products and Apparel (22 and 23)	4
Lumber Products (24)	0
Furniture (25)	0
Paper and Allied Products (26)	5
Printing and Publishing (Except Newspapers) (27)	1
Chemicals (Except Drugs) (28)	25
Drugs (283)	15
Petroleum Refining (29)	9
Rubber and Misc. Plastics (30)	5
Leather Products and Footwear (31)	1
Stone, Glass, and Clay Products (32)	7
Primary Metals (33)	8
Fabricated Metal Products (34)	10
Nonelectrical Machinery (35)	20
Electrical Machinery and Equipment (36)	19
Transportation Equipment (37)	18
Instruments (38)	5
Miscellaneous Manufactured Products (39)	2
Total	187

*Based on 1967 data.

Source: Raymond Vernon, Sovereignty at Bay (Basic Books, 1971), Table 1-2.

A national effort to block the transmission of knowledge from U.S. citizens to foreigners, to be at all
efficacious, would require a state policing apparatus
which probably would be intolerable to the nation's
citizenry.

Moreover, the power of technology transfer by U.S.
multinational firms to affect the domestic economy is
often overstated. It has been claimed, for example,
that such technology transfer is responsible for the
ills of the domestic apparel and footwear industries
(17). Yet, an examination of U.S. multinationals by
industry reveals that such firms by and large simply
do not operate in these industries. (See Table 2.)
Likewise, it has been claimed that U.S. firms are
transferring overseas their latest and best production
technologies and techniques. The detailed study by
Vernon and Davidson previoulsy cited, however, fails
to find significant evidence of such an occurrence.

- Slowdown in Technological Innovation. It is felt by
  some analysts that there has been a slowdown in the
  rate of technological innovation during the past ten
  or so years, in the sense that this rate has been below that which prevailed during most of the post-World
  War II era. Tangible evidence that a slowdown has
  occurred is actually quite slim. One fact in this
  direction is that the rate of measured increases in
  labor productivity has diminished during the past
  decade, although efforts to measure the reasons why
  this has occurred fail to point to reduced technological innovation as a dominant causal element (18). The
  total number of patents issued annually in the United
  States has declined somewhat since 1971, a fact suggestive of a slowdown in technological innovation, although it may also be that U.S. corporations are choosing to patent a lower percentage of new technologies.
  (See Table 3.) Most additional evidence used to demonstrate a slowdown has been anecdotal in nature (19).

The major effect of major technological slowdown on
U.S. trade would be that the nation would fail to renew its comparative advantage in the creation of new
products embodying advanced technologies. This is not
to say that the nation would stop exporting these goods
altogether. When a product reaches full maturity and
its technology becomes widely diffused, comparative
advantage in its production will be determined by non-
technological determinants such as relative factor
costs. The United States thus might in fact retain a

Table 3. U.S. Patents Granted, to U.S. and to Non-U.S.
Nationals, 1960-1977.

Year	All U.S. Patents Granted	U.S. Nationals	Non-U.S. Nationals
1960	47,170	39,472	7,698
1962	55,691	45,579	10,112
1964	47,379	38,411	8,964
1966	68,408	54,636	13,772
1968	59,103	49,783	13,320
1970	64,432	47,077	17,355
1971	78,320	55,979	22,341
1972	74,813	51,519	23,294
1973	74,148	51,509	22,639
1974	76,281	50,648	25,633
1975	72,029	46,731	25,298
1976	70,223	44,281	29,942
1977	65,218	41,452	23,766

Source: National Science Foundation, Science Indicators,
1968 (U.S. GPO, 1979), Table 4-17; from U.S. Patent and
Trademark Office Data.

comparative advantage in technology-intensive goods
based on factor costs. Alternatively, the nation might
retain some advantage based on product differentiation.
What would be lost would be that component of comparative advantage based on constant innovation of new technologies, and with this probably also would be lost the
ability of U.S. firms to capture rents or overseas sales
of recently innovated products. As will be demonstrated
in the next sub-section of this paper, however, there is
little evidence that such losses have yet occurred.

If a slowdown has indeed occurred, it has not made
itself felt in all sectors of the economy. In the
highly sophisticated semiconductor industry, for example, advances in technology have resulted since 1970
in almost a one-hundred fold decrease in the cost of
producing hand-held calculators, quartz timepieces,
and other products utilizing microelectronic components.
In the automotive industry, the rate of new product
innovation has noticeably accelerated in the 1970's.
It has been suggested by some forecasters that new
technologies in the life sciences which were still
largely in the laboratories during the 1970's will
become the basis for new commercial industries during
the 1980's. Although it may still be premature to say
so with confidence, it is possible that the 1970's
will be remembered as a highly productive decade,
contrary to present conventional wisdom.

C. Is There Evidence Indicating a Loss of U.S. Comparative Advantage Based upon Technological Innovation?

However persuasive one might find the arguments of the
previous section that the relative position of the United
States as an innovator of new technology should be on the decline, the available evidence does not show a major impact on
U.S. trade of such a decline. U.S. Department of Commerce
data, for example, indicates that the nation's trade balance
(excess of exports over imports) of technology-intensive
products has been positive throughout the 1970's (See Table 4).
Likewise, on a bilateral basis, the United States has main-
tained a positive balance of trade in these goods with each
of its major trading partners except West Germany and Japan.
(See Tables 5A and 5B.)

It is difficult, however, to conclude definitively that
these positive trade balances are a consequence of a continuing
commanding lead of the nation in the creation of new technology.
Alternative explanations can be advanced. The realignment of
currency parities in the early 1970's, for example, almost
surely has made U.S. exports more price competitive than they

Table 4. U.S. Trade Balance in "R&D-Intensive" Manufactured Products, 1960-1977 (in billions of dollars).

Year	Exports	Imports	Balance	Imports-Exports
1960	7.60	1.71	5.89	0.23
1962	8.72	2.00	6.72	0.23
1964	10.27	2.30	7.97	0.22
1966	12.17	4.18	8.00	0.34
1968	19.31	5.54	9.78	0.36
1970	19.27	7.55	11.72	0.39
1971	20.23	8.50	11.73	0.42
1972	22.00	10.99	11.01	0.50
1973	29.09	13.99	15.01	0.48
1974	41.11	17.24	23.87	0.42
1975	46.44	17.10	29.34	0.36
1976	50.83	21.87	28.96	0.43
1977	53.17	25.54	27.63	0.48

Note: Balance may not exactly equal exports minus imports due to rounding.

Source: National Science Foundation, Science Indicators 1978 (U.S. GPO, 1979), Table 1.21; from U.S. Department of Commerce data.

Table 5A. U.S. Trade Balance with Selected Areas and Nations, 1971-1977, in "R&D-Intensive" Manufactured Products (in billions of dollars).

Area or Nation	1971	1972	1973	1974	1975	1976	1977
Developing Nations	5.09	5.28	6.64	10.66	14.73	16.05	16.01
Western Europe	3.60	3.09	4.13	5.98	6.70	7.06	6.92
West Germany	0.19	-0.06	-0.21	-0.20	0.06	0.06	-0.07
Canada	1.87	2.33	3.00	4.24	4.83	4.73	4.53
Japan	-0.52	-0.97	-0.85	-0.55	-1.02	-2.75	-3.46

Source: National Science Foundation, Science Indicators, 1978 (U.S. GPO, 1979), Table 1-23; from U.S. Department of Commerce data.

Table 5B. Ratio of Imports to Exports of "R&D-Intensive" Manufactured Products, 1971-1977, with Selected Areas and Nations (in billions of dollars).

Area or Nation	1971	1972	1973	1974	1975	1976	1977
Developing Nations	0.15	0.23	0.26	0.24	0.17	0.20	0.24
Western Europe	0.48	0.58	0.57	0.52	0.51	0.52	0.56
West Germany	0.85	1.04	1.13	1.10	0.97	0.98	1.03
Canada	0.52	0.50	0.48	0.43	0.41	0.46	0.51
Japan	1.34	1.59	1.38	1.18	1.43	2.06	2.24

Source: Same as Table 5A.

were during the last years of fixed exchange rates. Thus U.S.
trade performance in goods categorized by the Commerce Department as "technology-intensive" may have been as much bolstered
by the decline of the dollar as by a continuing "technology
gap." It is also possible that U.S. factor endowments would
cause the nation to retain a comparative advantage in the manufacture of these goods even if no "technology gap" existed at
all. Whatever the explanation, however, it would appear that
for the moment the nation's performance in international trade
of "technology intensive" goods remains strong.

A study published by Bela Balassa in late 1977 examined
changes in export competitiveness of industries of several
advanced nations for the years 1963-1971 (20). The measure
used was a nation's share of world exports in an industry
divided by that nation's share of total world exports. Balassa
found that during this time period the United States actually
increased its "revealed" advantage in most categories of re-
search-intensive goods, the major exceptions being in medical
and pharmaceutical products and in certain categories of
electrical goods. To be sure, these results are based on old
data, but they reinforce the point that as of yet, there is
no strong evidence of a major decline of U.S. competitiveness
in world trade of goods embodying advanced technologies.

It is interesting to note that U.S. imports of technology-
intensive goods as a ratio of exports of these goods grew
steadily, and significantly, from 1960 through 1972. (Refer
to Table 4.) From 1972 through 1975, however, this ratio
dropped, reflecting perhaps both slow economic growth within
the United States (which tends to be reflected in lower
growth of imports) and increased price competitiveness of U.S.
manufactures. From 1976 through 1977, this ratio again grew
somewhat. This was caused almost entirely by rapid growth of
imports, rather than reduced growth of exports. The import
growth resulted from rapid growth of the U.S. economy relative
to Western Europe.

II. Choices for the United States

It was noted earlier in this paper that the changing
world economic environment in general, and higher oil prices
in particular, may have reduced the degree to which U.S. comparative advantage in international trade is determined by the
nation's lead in technological innovation. The changing circumstances require adjustment in the U.S. economy. Adjustment
implies transfer of resources from less to more productive
uses, and this may be accomplished in two ways: intrasectoral
adjustment, whereby resources are transferred within a sector
so as to rationalize plant and equipment and/or modify end

products, and intersectoral adjustment, whereby resources are
transferred out of industries in decline and into those whose
futures are brighter. It has also already been noted that the
adjustment can be facilitated by means of the import of foreign
technologies. The import can be accomplished via licensing
or imitation of foreign technologies and by import of foreign
durable goods embodying these technologies. In addition to
using these technologies directly, U.S. industry will be induced by foreign competition to develop its own technologies
more suited to the changing times than those currently in use.

At the beginning of this decade, the U.S. Government is
finding itself virtually under siege from industries and organized labor groups seeking relief from import competition.
Traditionally protectionist industries such as textiles, apparel, shoes, and steel have been joined by the automotive
producers and even to a degree by the very high technology
semiconductor industry. The case against protectionism has
already been covered in this paper: not only do protectionist
policies inhibit the flow of needed new technologies into the
economy, they impose an implicit tax on the users of the imported products as well.

Nonetheless, imports do displace domestic workers who
produce similar products, and the fact that new jobs are being
created in other industries is of scant comfort to the worker
whose job is threatened by imports. If imports in a particular sector rise more quickly than adjustment can take place,
net unemployment will ensue. Some industries, notably the
automotive industry, believe that they can successfully adjust
intrasectorally to meet foreign competition, but need "breath-
ing time" for the adjustment to take place. Some protection
of those industries in which imports have surged in recent
times is probably justifiable, so long as the protection is
short-lived and not renewed once its term as originally set
has run out.

Unfortunately, once granted, protection often becomes
self-perpetuating. And when perpetuated, the protection does
the economy irreparable harm. Resources are locked into inefficient uses, the adoption of new technologies is retarded,
and consumers are forced to pay higher prices. The result in
the long run is higher prices, lower productivity, and, ultimately, reduced economic growth and lower standards of living.

It has been noted that adjustment takes place most easily
during periods of rapid economic growth and, conversely, that
the barriers to adjustment rise during periods of slow growth
(21). But it is also true that lack of adjustment can be a re-
tardant to growth. Thus, one of the most important tasks

facing U.S. policymakers in the near future will be to develop rational adjustment policies. The choices won't be easy. A certain amount of structural unemployment may have to be tolerated to ensure that industries affected by rising imports do in fact adjust, and pressures to overprotect such industries are likely to grow.

The case probably can be made that it is worthwhile to grant temporary protection to those industries which have the potential to adjust intrasectorally and not to protect at all those industries in which the nation holds no comparative advantage. Unfortunately, political pressures are such that industries in the latter category often receive the most protection on a protracted basis.

The automotive industry could very well set a precedent for cases to come of industries which urgently must adjust but are potentially highly competitive. The tradeoffs in this industry are fairly straightforward: protection of domestic jobs in the sector versus continued importation of vehicles which, at the moment at least, are preferred by a significant number of U.S. consumers over domestically produced cars for a number of reasons, one of which, fuel efficiency, is congruent with the national goal of reduced dependence on foreign oil. There is little doubt that given time and resolve the U.S. industry can efficiently produce competitive products. How much and what type of protection, if any, should be afforded the industry and for how long has become an urgent matter.

Exactly what adjustment policies the United States Government ought to pursue (beyond not adopting protectionist policies) is a difficult issue. Adjustment in many sectors implies a higher than normal rate of new capital investment, and policies to give firms incentives to invest (such as accelerated depreciation) doubtlessly are warranted. At the present time, the Government provides transfer payments to both workers and firms whose fortunes have been adversely affected by imports, through the Labor and Commerce Departments' trade adjustment assistance programs. One problem with such programs is that they may tend to impede adjustment by encouraging labor and capital to remain immobile rather than providing an incentive to change. In spite of this, some level of transfer payment can be justified on grounds of social equity. How much transfer payment is warranted and for how long is problematic, and the trend seems to be generally in the direction of increase.

Whatever the policies of the U.S. Government, the brunt of the burden of adjustment will fall on the private industrial sector. In order to remain competitive internationally,

many individual firms will have to reorient their internal
strategies. The need for reorientation is likely to be
especially great for firms whose success in export markets
has rested upon unique product innovations. As international
competition for these products grows, and as the products themselves mature, the bases for success in world markets increasingly become price, service, and quality assurance rather than
the uniqueness of the product itself. Thus, firms whose strategies revolved around new product development without much
emphasis on internal efficiency or customer service will find
themselves facing new challenges for which their internal
organizations may not be well suited to respond.

While these last remarks apply particularly to firms
operating in the so-called "high technology" industries, and
small to medium-sized firms especially, it is safe to say that
much of American industry ought to pay greater attention to
internal efficiency and quality assurance. Even in the automotive sector, in which U.S. firms justifiably can point with
pride to a record of manufacturing efficiency and product
reliability, it is evident that some of the recent success of
Japanese imports is due to even greater levels of efficiency
and product reliability on the part of Japanese producers.

To achieve both greater efficiency and product reliability,
greater attention must be paid by much of U.S. Industry to
adopting the latest and the best technologies at the level of
the individual manufacturing plant. It is a simple fact today
that much of the latest and best plant technology is held
abroad. It has been noted, for example, that Japanese manufacturers are generally ahead of U.S. firms in developing and
applying robotic techniques for the automation of assembly
operations (22). In some cases, U.S. industry may even be
laggard in adopting U.S.-developed process technologies. Discussions between this author and executives of one large U.S.
computer manufacturer have revealed that in some industries,
especially those characterized as "mature," foreign firms much
more readily adopt the latest computerized process control
technologies than do U.S. firms (although, it must be noted,
this situation apparently is changing).

It would, of course, be to undersell U.S. industry to
claim that some of this adjustment is not already taking
place. The rising percentage of U.S. patents held by foreigners (see Table 3) may be indicative of increasing adoption of
needed foreign technologies by U.S. firms. Rather than a
cause for alarm as suggested by some analysts, the trends
revealed in the patent statistics may signal the beginning of
a needed process of adjustment.

Product differentiation also is likely to be an increasingly important determinant of U.S. competitiveness in world trade. To be successful in a foreign market, a product must often be adapted or modified to meet the particular requirements and tastes of that market's consumers. Thus, the type of product differentiation that is called for is adaptation of U.S.-made products to the special characteristics of foreign markets. Persons familiar with Japanese export policy note that one element in Japan's success has been adaptation of Japanese-made products to local market conditions, and this aspect of that nation's policy is well worth emulation.

Adjustment cannot take place speedily in the face of organized worker resistance, and the inflexibility of trade unions is frequently cited as one reason for the declining status of Great Britain as an industrial power. At least one study shows, however, that unionized U.S. workers are not as opposed to technical change as is sometimes supposed, especially in situations where technological adaptation is vital to preservation of jobs (23). Trade union flexibility is likely to be reduced, however, if union leadership perceives that lobbying for protectionist policies in Washington can preserve more jobs with less effort than can adaptation at the level of the plant. Thus, the ability of the Federal Government to stand up to protectionist drives on the part of organized labor is underscored as an important determinant of the ability of the U.S. economy to remain competitive.

The process of adjustment will prove to be a boon overall to some sectors of the economy, as well as posing hardships for other sectors and headaches for public policy formulators. Redesign of U.S. automobiles and the ensuing retooling of the industry's plant and equipment, for example, has stimulated new economic activity in several industrial centers of the Midwest which had become pockets of depression during the 1970's. (See "Redesign of American Cars a Boon to Midwest Industry," New York Times, March 3, 1980, page 1.) Capital goods producers throughout the nation will find their back-orders growing, albeit that some of the demand will be for redesigned tools and capital goods embodying new and different technologies which old-line producers may find difficult to fulfill without themselves undergoing a period of adjustment.

On the international side, the United States Government must continue to press to open new markets for U.S.-made goods and to work to ensure that existing markets remain open. The job is complicated by protectionist pressures abroad which are mounting just as they are in the United States. It is clear that U.S. markets cannot remain open to foreign goods if foreign markets become closed to U.S. goods.

Possibly the largest overseas market which has been largely closed to U.S. manufacturers of civilian goods is that created by foreign government and state-owned enterprise procurement. This market accounts for as much as a third or more of gross domestic product in some western European nations. To a large degree, "buy American" provisions of state, local, and federal laws governing procurement in the United States have also effectively closed the U.S. government procurement market to foreign suppliers. Much government procurement, both in the United States and abroad, is of high technology goods, and an opening of overseas markets would be especially of benefit to U.S. producers of these goods. The Government Procurement Code, enacted as part of the Multilateral Tariff Negotiations of 1979, is a first step towards opening these markets to international competition whereby governments will open portions of their procurement markets to foreign suppliers on a reciprocal basis. This code is, however, but a first step in the right direction. The opening of procurement markets represents the greatest unexploited opportunity existing for further trade expansion among the major industrial states of Western Europe, North America, and Japan, an expansion which could ease adjustment problems in all three areas. It is an opportunity which should continue to be vigorously pursued.

References

(1)   For a review of this literature, see Graham (1979).

(2)   See Vernon (1966), (1971), 2d (1974). For a review of the "key ideas," see Wells (1972).

(3)   See Wells (1969), Stobaugh (1971), and Tilton (1971).

(4)   See Kennedy (1964).

(5)   See Davidson (1976).

(6)   See Utterback (1974), Abernathy and Townsend (1975), Utterback and Abernathy (1975), and Utterback (1979).

(7)   The point is developed by Johnson (1970).

(8)   See Wells (1972), op. cit.

(9)   See Vernon (1971), op. cit., Chapter 3.

(10)   See Vernon (1979), Graham (1979), op. cit.

(11)   See Klein (1977) and (1979).

(12)   See Linder (1961) and Hufbauer (1970).

(13)   See Organization for Economic Cooperation and Development (1979).

(14)   See Vernon and Davidson (1978).

(15)   Mansfield et al.(1979) calculate that, for a same of firms, upwards of 30% of the total return on $ investment in new technology comes from foreign sales.

(16)   See Kindleberger (1974) for an account of British efforts to "contain" technology.

(17)   See Goldfinger (1974), which contains an explanation of what has become a point of view of the AFL-CIO on technology transfer.

(18)   See Denison (1979).

(19)   For an account of some such evidence, see National Academy of Engineering (1978).

(20)   See Balassa (1977).

(21)   See OECD (1979A).

(22)   See Norman Macrae, "Must Japan Slow," The Economist 274, February 1980, after page 60.

(23)   See Bourdon (1979).

Bibliography

W. J. Abernathy and P. L. Townsend (1975), "Technology, Productivity, and Process Change, "Technological Forecasting and Social Change" 7, pp. 379-396.

Bela Balassa (1977), "Revealed Comparative Advantage Revisited: An Analysis of Relative Export Shares of the Industrial Countries, 1953-1971," Manchester School of Economics and Social Studies, pp. 327-344.

Clinton C. Bourdon (1979), "Labor, Productivity, and Technological Innovation: From Automation Scare to Productivity Decline," in C. T. Hill and J. M. Utterback, editors, Technological Innovation for a Dynamic Economy (New York: Pergamon Press) pp. 222-254.

William H. Davidson (1976), "Patterns of Factor-Saving Innovation in the Industrialized World," European Economic Review 8, pp. 207–218.

Edward F. Denison (1979), Accounting for Slower Economic Growth in the 1970's: The United States in the 1970's (Washington, D.C.: The Brookings Institution).

Nat Goldfinger (1974). "A Labor View of Foreign Investment and Trade Issues," in R. E. Baldwin and J. B. Richardson, editors, International Trade and Finance (Boston: Little, Brown, and Co.).

Edward M. Graham (1979), "Technological Innovation and the Dynamics of U.S. Comparative Advantage in International Trade," in Hill and Utterback, editors, Technological Innovation for a Dynamic Economy, pp. 118–160.

Gary C. Hufbauer (1970), "The Impact of National Characteristics and Technology on the Commodity Composition of Trade in Manufactured Goods," in R. Vernon, editor, The Technology Factor in International Trade (Cambridge, Mass: National Bureau of Economic Research).

Harry G. Johnson (1970), "The Efficiency and Welfare Implications of the International Corporation," in C. P. Kindleberger, editor, The International Corporation (Cambridge, Mass: The MIT Press).

Charles Kennedy (1964), "Induced Bias in Innovation and the Theory of Distribution," The Economic Journal 74, pp. 541–547.

Charles P. Kindleberger (1974), "An American Climacteric," Challenge 16, pp. 35–44.

Burton H. Klein (1977), Dynamic Economics (Cambridge, Mass: Harvard University Press).

_____ (1979), "The Slowdown in Productivity Advances: A Dynamic Explanation," in Hill and Utterback, editors, Technological Innovation for a Dynamic Economy, pp. 66–117.

Staffan B. Linder (1961), An Essay on Trade and Transformation (Stockholm: Almquist and Wiksell).

E. Mansfield, A. Romeo, and S. Wagner (1979), "Foreign Trade and U.S. Research and Development," Review of Economics and Statistics 61, pp. 49–57.

National Academy of Engineering (1978), Technology, Trade, and the U.S. Economy (Washington, D.C.: National Research Council).

Organization for Economic Cooperation and Development (1979), The Impact of the Newly Industrializing Countries on Production and Trade in Manufactures (Paris: OECD).

Organization for Economic Cooperation and Development (1979A), The Case for Positive Adjustment Policies (Paris: OECD).

Robert B. Stobaugh (1971), "The Neotechnology Account of International Trade," Journal of International Business Studies, pp. 43-64.

John E. Tilton (1971), International Diffusion of Technology: The Case of Semiconductors (Washington, D.C.: The Brookings Institution).

James M. Utterback (1974), "Innovation in Industry and the Diffusion of Technology, Science 183, pp. 620-626.

_____ (1979), "The Dynamics of Product and Process Innovation in Industry," in Hill and Utterback, editors, Technological Innovation for a Dynamic Economy, pp. 40-65.

James M. Utterback and William J. Abernathy (1975), "A Dynamic Model of Product and Process Innovation," Omega 3, pp. 639-656.

Raymond Vernon (1966), "International Investment and International Trade in the Product Cycle," Quarterly Journal of Economics 80, pp. 190-207.

_____ (1971), Sovereignty at Bay: The Multinational Spread of U.S. Enterprises (New York: Basic Books).

_____ (1974), "The Location of Economic Activity," in John H. Dunning, editor, Economic Analysis and the Multinational Enterprise (London: George Allen and Unwin).

Louis T. Wells (1969), "Test of a Product Cycle Model of International Trade: U.S. Exports of Consumer Durables," Quarterly Journal of Economics 83, pp. 152-162.

_____ (1972), "International Trade: The Product Life Cycle Approach," in L. T. Wells, editor, The Product Life Cycle and International Trade (Cambridge, Mass: Harvard University Press).

2. Industrial Innovation and Government Policy

I am very pleased to have the opportunity to discuss the
topic of innovation -- the means by which new products and
processes are brought into commercial use. It is an issue
which has attracted close attention in recent years in the
executive branch and in the Congress and the general public.
It is important to consider some of the ways in which the
Carter administration confronted this important issue. My
paper will be divided into three parts. First, I will
describe the background of the issue and the history sur-
rounding the administration's efforts. Second, I will
describe some of the initiatives that the Carter administration considered. Finally, I will turn to the important
issues that remain ahead.

By way of background, let me describe how we in the
office of the President's science and technology adviser
first became concerned with the issue. "Innovation" is itself
an ill-defined process, and there is no single indicator that
can give a measure of its health and vitality. However, there
were several coupled trends that we observed in early 1978
that gave us cause for concern. I might add that our concerns
were shared by many others, in particular, representatives of
the Department of Commerce.

First, we looked to the inputs of the innovation process.
Research and development, as a percentage of our Gross
National Product, had been steadily declining for nearly ten
years. Even more disturbing, was the steep decline in the
same period in the support of basic research -- the investment in the long-term work that leads to whole new industries.
And these trends were supplemented by the views of industrial
managers who sensed that their firms were focusing their
research on incremental product improvement rather than major
advances. Capital investment was also growing much more
slowly in the United States than elsewhere, and over a

seven-year period there was a marked decline in the startup
of new high-technology companies.

On the output side, we also saw reason for concern. The
trade balance in R&D-intensive manufactured products -- the
products where our technological edge can compensate for
increased labor costs -- was suffering declines, and Japan
was posing a particular threat. The productivity growth in
the United States was less than that of any western nation.
(Indeed, in the last year, productivity growth in the United
States has been negative.) And the 1971-1977 period saw a 26
percent decline in United States patents granted to U.S. citizens.

Finally, regardless of the significance of these trends,
efforts to augment our innovative activities were clearly
worthwhile. A wide range of studies has revealed their innovation could be an important component in addressing several of
the challenges confronting our nation. Among these were the
need to increase productivity, to foster economic growth, and
to decrease inflation. Indeed, economists believe science
and technology is the major factor in economic and productivity growth. Moreover, innovation clearly is essential in
developing new technologies for energy supply and conserva-
tion, in improving health care, and providing technologies to
increase food production in an increasingly hungry world.

We concluded that the issue was worthy of closer examination and the development of policy at the presidential
level. This was an accomplishment in itself. We tend to
think of innovation as an American birthright -- we pride
ourselves on Yankee ingenuity (at least north of the Mason-
Dixon Line.) But we cannot afford to be complacent. Other
countries were the paramount world innovators in the 19th
century, and today we largely look elsewhere for innovative
skills.

We were then confronted with the problem of how best to
address the issue. A moment's reflection will reveal that
innovation is affected by the business of nearly every agency
in Washington. Regulatory policy, antitrust policy, the
funding and mechanisms for funding research and development,
procurement, small business policy, and labor policy, for
example, all play a part. Unfortunately, innovation had never
been a paramount factor in the thinking of any agency. Thus,
it was necessary to bring together a wide variety of agencies
to discuss the issue and to develop a program for the President.

The effort was therefore an interagency one, undertaken
under the leadership of the Department of Commerce. Because

it was our intent to affect behavior in the private sector --
it was our hope to make the private sector more innovative --
we started out the study by forming an extensive advisory
committee. The group was formed from members of large and
small business, labor, public interest groups, and the academic
community. We sought guidance as to how the government could
best revise its policies so as either to minimize its interference with or to enhance the process of innovation in the
private sector. The output from this advisory committee was
then fed into an interagency group, which responded to their
recommendations. Ultimately, the recommendations of the
interagency group worked their way into the White House, and
the presidential statement was issued in late October 1979.

Let me turn now to some of the initiatives that President
Carter announced. I must caution at the outset that innovation is an exceptionally subtle process, and there is no
one initiative or cluster of initiatives that by themselves
will guarantee the enhancement of innovation in the United
States. As a result, a broad spectrum of initiatives must be
undertaken and adjusted over time. A sensible program cannot
stand on actions in a single area.

Let me demonstrate this fact by a simple example. Consider the process by which a new pharmaceutical to treat a
previously untreatable disease is brought to market. The
process might well start with research, perhaps basic research,
funded by the government through the National Institutes of
Health. The government or the industry might fund research
into the character of the biochemical processes that affect
the course of the disease, for example. These results might
in turn lead to further research by a pharmaceutical company
to develop a compound for treatment. Even if we suppose that
this research is productive, there is no assurance that the
product will ever come into the market. Several other things
must happen. First the pharmaceutical company must engage in
an extended regulatory process with the Food and Drug
Administration to assure the safety of the drug. As you know,
this may take years and may involve very expensive programs
involving laboratory animals. Second, the company must amass
the capital that is required to perform this testing and to
go into production. Finally, there are obvious questions of
patents. Without adequate patent protection, the company may
be unwilling to engage in any of this previous investment
because of the fear that a competitor will copy and reap the
benefits without incurring any of the costs. (Indeed, as you
might imagine, there is a complicated interaction between the
limited term of the patent and the length of the regulatory
process.) In this simple example, innovation is affected by
research funded by various sources, regulatory policy,

economic policy, and patent policy -- any one of which could
thwart the innovation from occurring. Imagine the difficulties when sales to the government, multiple firms, or small
business are involved. In sum then, innovation is extremely
complicated.

Because of time constraints, I will not have an opportunity to discuss all of the 32 initiatives that the President
announced.* It would take me the entire session to do justice
to the topic. Let me turn now to a few of the major initiatives and discuss them briefly.

1. <u>Industry-University Cooperation</u>. A particularly
important portion of our federal investment in R&D, about
$5 billion a year, goes to the support of basic research.
This is the research for which the private sector does not
have adequate incentive to invest by itself. For the most
part, the results are unappropriable and cannot be patented.
Moreover, the benefits are likely to be achieved only in the
long term -- a fact that discourages investment by business
managers. And finally, an individual basic research project
is risky in the sense that there is a large measure of chance
in whether the research will lead to commercial payoffs or
not. In the aggregate, basic research is not risky, but, with
the exception of a few industrial giants, no one firm is
likely to be able to support a sufficient portfolio of basic
research to be able to achieve the aggregate benefits. As a
result of these facts, there is a clear public need to fund
basic research. Indeed, in recognition of these facts, the
Carter administration increased the federal support for such
work over the past three terms by nearly 35 percent.

Much of this work is appropriately done in universities --
the academic environment, the attitudes toward science, the
freedom afforded researchers all make university scientists
particularly productive. But this creates a dilemma. The
ultimate application of such work in the economy requires the
involvement of industry. There, thus, is an important need for
a linkage between the industry and the university communities.
Unfortunately, these linkages were strained or sundered in the
Vietnam era and are only slowly starting to be regenerated.

In order to speed this process, the President has
announced the significant augmentation of an industry-
university cooperative research. This program at the National
Science Foundation will grow to nearly $15 million in the
Fiscal year 1981 and move to an eventual government-wide goal

---

*A message to the Congress and a Fact Sheet describing the
initiatives are attached as Appendices A and B.

of nearly $150 million. The program provides special incen-
tives for research proposals that have joint investigators --
one from industry and one from the university. The proposals
go through the normal peer-review process to assure that they
have scientific merit, but they benefit from a protected pot
of money. This program is properly targeted because it
builds the linkages at the appropriate level -- from university
scientist to industrial scientist. It is warmly endorsed by
both sectors.

Let me add in passing, the administration has also
embarked on a few cooperative ventures with particular industrial sectors. One of these is a $1 billion program to per-
form basic research on automotive technology. The program
will be jointly funded by the government and by the auto
industry, and the performers, again, will be both universities
and industrial laboratories. The administration is also
launching in its FY 1981 budget a cooperative program with the
oil industry to investigate the ocean margins -- deep water
regions on the edge of the continental shelf that have never
been explored, but which are believed to be important in the
geophysical history of the earth. Although the program is
one driven by scientific objectives, the resource potential
of the margins is vast and should be assessed. Moreover,
the program will develop needed technology to drill in deep
water.

2. <u>Antitrust</u>. The antitrust laws enjoy a very complicated relationship to the innovation process. On the one hand,
competition can provide the incentive to innovate and thus can
be an important stimulus to the innovation process. On the
other hand, however, there occasionally are circumstances in
which cooperation among firms is necessary for advance. For
example, clusters of small firms may not have sufficient
available funds to each maintain independent research capabilities and by combining, they can achieve economies of scale
and support research groups of the necessary "critical mass"
for productive work. Unfortunately, the antitrust laws are
commonly understood as forbidding such cooperative ventures.

One of the most personally rewarding aspects of the
innovation study was the dialogue that we were able to establish with the antitrust division of the Justice Department.
There was a mutual education process that went on -- we
became sensitized to the department's legitimate antitrust concerns, and I believe the department learned about research and
development and the innovation process. The product of this
effort will be available to the public shortly. The department will issue a guide to cooperative activity in research
and development that should provide substantial help to

businessmen who contemplate cooperative ventures.* The issuance of such a guide is a noteworthy occurrence, as only two
similar guides have been issued by the department in the past.
We trust that the dialogue we have started will continue with
industry in the future.

3. Small Business. A variety of independent observers
have noted the important role that small businesses have
played in the innovation process. Major breakthroughs often
require an entrepreneur in a small business who is willing to
take the substantial risks and to create a market where one
has not existed before. In light of the importance of the
small entrepreneur, the President has announced a cluster of
programs that have special focus on small business. Chief
among these is a program at the National Science Foundation
to support creative, high-risk, potentially high-reward
research and concept development by small business. This program is itself a new approach and is widely acclaimed by the
small business community. In fact, venture capitalists have
been interested in providing financing to the companies who
have participated. We have set a goal of $150 million in the
program.

4. The Patent System. There are a number of initiatives
that the President has announced to reform and streamline the
patent system. Patents pay a particularly important role in
innovation. Indeed, that is their purpose. They provide the
inventor with an incentive. They also stimulate a firm to
make the investment that is required to bring an invention to
market. Usually copying is much cheaper than inventing, and
thus the monopoly provided by the patent system is essential.
Finally, patents provide a method of the disclosure of
information. The President announced several initiatives to
better achieve these objectives:

● The administration has submitted a bill to the Congress
   to allow the voluntary re-examination of patents. This
   will enable the patent office to deal with prior art of
   which it was unaware at the time of patent issuance and
   should greatly improve the reliability of patent deter-
   minations in the courts.

● The administration has again urged the Congress to establish a single court to deal with patent appeals. The
   court would establish nationwide uniformity in patent law,
   make litigation more predictable, and eliminate expensive
   forum shopping.

---

* "Antitrust Guide Concerning Research Joint Ventures." U.S.
Department of Justice, Antitrust Division, November 1980.

● The administration will work with the Congress to resolve
  the nagging issue concerning allocation of patent rights
  arising from federal sponsorship. The departments and
  agencies of government are currently guided by some
  twenty different statutes embodying a wide variety of
  schemes for allocating patent rights. For the most
  part, these statutes provide that the government shall
  retain title to the patents, with unrestricted licenses
  available to the general public. Unfortunately, unless
  an entrepreneur has exclusive rights, it is unlikely that
  he will make the investment to bring a patented object to
  market. Thus, the paradox that making the patent available to all results in the unavailability of the patented
  invention. In light of this fact, the President has proposed that the contractor be allowed to obtain exclusive
  rights to patents arising from his work in fields in which
  he commits himself to commercialize the invention. In
  the case of universities and small businesses, the administration would allow the retention of patent title.

● The President also announced his intention to commence
  the substantial upgrading of the patent office. We are
  very much aware of the fact that funding in the patent
  office has in the past been inadequate. There is a need
  for funds for improved filing and classification systems,
  and I believe we will see more efforts to use modern
  technology to speed or improve the patent process.

5. Regulation. Regulation can both spur innovation and
inhibit it. In some cases, such as in automotive emission
devices, the regulatory system was used to force change. In
other instances, regulation can cause industry to focus on
protecting its existing product lines, rather than creating
new ones. Moreover, the uncertainty created by some regulatory schemes can discourage the long-term commitment that
major innovation requires.

The administration has embarked on many efforts to eliminate needless regulation (as in the case of airlines) and to
streamline and rationalize needed regulation. We are seeking
to assure that regulation has an adequate scientific basis
and, where appropriate, to assure that cost-impact analysis
is undertaken. Other efforts deal with coordination of regulatory activities, the use of performance goals that allow
and encourage the industry to be innovative in meeting the
goals, and the establishment of a forecast of upcoming regulation so industry has the time to develop new technology.

I could go on at some length with a large number of other
initiatives that the President has announced, such as those

dealing with labor or procurement, but I will not. The important point, and one which I seek to emphasize here, is the
need to experiment with a wide range of policies. No single
approach is equally important for business of all sizes, or
for various industrial sectors. Different and novel approaches
are necessary.

I must also mention the important step we did not take.
If one speaks to industry about the important problems which
confront them, one would speak chiefly of two major issues,
regulation and taxes. Although we have made significant
advances in the regulatory area -- and will continue to strive
to improve the system -- the President, in announcing his
initiatives, chose not to adopt tax measures. He did state
that in examining his fiscal policies in the future, he would
again consider the issue. Once we have established fiscal
discipline and brought inflation under control, I expect tax
measures to encourage innovation, capital formation, and productivity.

Finally, I must state the obvious: no one study or the
initiatives launched as a result of it can transform the
entire economy. The issue is not behind us. There must be a
continuing commitment to encourage innovation in the country.
In fact, encouraging innovation, increasing productivity, and
stimulating investment will, I believe, become commonplace
elements of the policy formation in all of the agencies in the
years ahead. We are starting to see this happen. But we have
only seen an early skirmish in what must be a continuing
battle to maintain the technological strength of the American
economy.

APPENDIX A

President Carter's Message to Congress, October 31, 1979

Office of the White House Press Secretary

THE WHITE HOUSE

TO THE CONGRESS OF THE UNITED STATES:

Industrial innovation -- the development and commercialization of new products and processes -- is an essential
element of a strong and growing American economy. It helps
ensure economic vitality, improved productivity, international
competitiveness, job creation, and an improved quality of life
for every American. Further, industrial innovation is necessary if we are to solve some of the Nation's most pressing
problems -- reducing inflation, providing new energy supplies
and better conserving existing supplies, ensuring adequate
food for the world's population, protecting the environment
and our natural resources, and improving health care.

Our Nation's history is filled with a rich tradition of
industrial innovation. America has been the world leader in
developing new products, new processes, and new technologies,
and in ensuring their wide dissemination and use. We are
still the world's leader. But our products are meeting growing competition from abroad. Many of the world's leading
industrial countries are now attempting to develop a competitive advantage through the use of industrial innovation.
This is a challenge we cannot afford to ignore any longer.
To respond to this challenge, we must develop our own policies
for fostering the Nation's competitive capability and entrepreneurial spirit in the decades ahead. This Message
represents an important first step in that direction.

I am today announcing measures which will help ensure our
country's continued role as the world leader in industrial
innovation. These initiatives address nine critical areas:

●   Enhancing the Transfer of Information
●   Increasing Technical Knowledge
●   Strengthening the Patent System
●   Clarifying Anti-trust Policy
●   Fostering the Development of Small Innovative Firms
●   Opening Federal Procurement to Innovations
●   Improving Our Regulatory System
●   Facilitating Labor Management Adjustment to
    Technical Change
●   Maintaining a Supportive Climate for Innovation.

INITIATIVES

1. Enhancing the Transfer of Information. Often, the information that underlies a technological advance is not known to companies capable of commercially developing that advance. I am therefore taking several actions to ease and encourage the flow of technical knowledge and information. These actions include establishing the Center for the Utilization of Federal Technology at the National Technical Information Service to improve the transfer of knowledge from Federal laboratories; and, through the State and Commerce Departments, increasing the availibility of technical information developed in foreign countries.

2. Increasing Technical Knowledge. We have already made significant efforts to assure an adequate investment in the basic research that will underlie future technical advances. This commitment is reflected in a 25 percent growth in funding during the first two years of my Administration. I am taking some additional steps that will increase Federal support for research and development:

First, I will establish a program to cooperate with industry in the advancement of generic technologies that underlie the operations of several industrial sectors. This activity will broaden the $50 million initiative I announced in May to further research in automotive research. Second, in order to help harness the scientific and technological strength of American universities, I have directed a significant enhancement in support of joint industry-university research proposals. This program will be modeled on a successful program at the National Science Foundation, and I have set a target of $150 million in Federal support for it.

3. Strengthening the Patent System. Patents can provide a vital incentive for innovation, but the patent process has become expensive, time-consuming, and unreliable. Each year, fewer patents are issued to Americans. At my direction, the Patent and Trademark Office will undertake a major effort to upgrade and modernize its processes, in order to restore the incentive to patent -- and ultimately develop -- inventions. I will also seek legislation to provide the Patent and Trademark Office with greater authority to re-examine patents already issued, thereby reducing the need for expensive time-consuming litigation over the validity of a patent.

For over thirty years the Federal agencies supporting research and development in industry and universities have had conflicting policies governing the disposition of pertinent rights resulting from that work. This confusion has

seriously inhibited the use of those patents in industry. To
remove that confusion and encourage the use of those patents
I will support uniform government patent legislation. That
legislation will provide exclusive licenses to contractors in
specific fields of use that they agree to commercialize and
will permit the government to license firms in other fields.
If the license fails to commercialize the inventories, the
government will retain the right to recapture those rights.
I will also support the retention of patent ownership by
small businesses and universities, the prime thrust of legislation now in Congress, in recognition of their special place
in our society.

4. Clarifying Anti-trust Policy. By spurring competition,
anti-trust policies can provide a stimulant to the development
of innovations. In some cases, however, such as in research,
industrial cooperation may have clear social and economic
benefits for the country. Unfortunately, our anti-trust laws
are often mistakenly viewed as preventing all cooperative
activity.

The Department of Justice, at my direction, will issue a
guide clearly explaining its position on collaboration among
firms in research, as part of a broader program of improved
communication with industry by the Justice Department and the
Federal Trade Commission. This statement will provide the
first uniform anti-trust guidance to industrial firms in the
area of cooperation in research.

5. Fostering the Development of Small Innovative Firms.
Small innovative firms have historically played an important
role in bringing new technologies into the marketplace. They
are also an important source of new jobs. Although many of
the initiatives in this Message will encourage such companies,
I will also implement several initiatives focused particularly
on small firms.

First, I propose the enhancement by $10 million of the
Small Business Innovation Research Program of the National
Science Foundation. This program supports creative, high-
risk, potentially high-reward research performed by small
business. Further, the National Science Foundation will
assist other agencies in implementing similar programs, with
total Federal support eventually reaching $150 million per
year.

Second, in order to experiment with ways to ease the
ability of small firms to obtain start-up capital, I will
help establish two Corporations For Innovation Development to
provide equity funding for firms that will develop and market

promising high-risk innovations. These not-for-profit firms
will be established with State or regional capital and the
Federal government will provide each with matching loan funds
up to $4 million.

6. <u>Opening Federal Procurement to Innovations</u>. The
Federal government is the Nation's largest single purchaser
of goods and services. Through its purchases, the Federal
government can influence the rate at which innovative products
enter the market.

For that reason, I am directing the Office of Federal
Procurement Policy to introduce procurement policies and
regulations that will remove barriers now inhibiting the
government from purchasing innovative products. Special
attention will be given to substituting performance for
design specifications and, wherever feasible, selection will
be on the basis of costs over the life of the item, rather
than merely the initial purchase price.

7. <u>Improving our Regulatory System</u>. During my Administration, I have already taken a number of actions to help
assure that regulation does not adversely affect innovation.
Working with the Congress, I have moved successfully toward
deregulation of airlines and other industries, and I expect
the pressure of competition to trigger innovative new ways to
cut costs and improve service. In environmental, health and
safety regulation, I have emphasized the use of cost-impact
analysis, where appropriate, to take account of the burdens
on industry in the regulatory process. To provide better
coordination between the regulatory agencies, I have created
the Regulatory Council, composed of the heads of 35 regulatory
agencies. This Council is working to reduce inconsistencies
and duplications among regulations, to eliminate needless
rule-making delays, to reduce paperwork, and to minimize the
cost of compliance.

I am today proposing additional steps to improve our
regulatory system. First, the Administrator of EPA will
intensify his efforts, wherever possible, to use performance
standards in regulations, specifying only the required goal,
rather than the means of achieving it. Second, all Executive
Branch environmental, health and safety regulatory agencies
will prepare a five-year forecast of their priorities and
concerns. This information will give industry the time to
develop compliance technology. Third, all administrators of
Federal executive agencies responsible for clearance of new
products will be directed to develop and implement an expedi-
ted process for projects having a strong innovative impact or

exceptional social benefit, and to do so without jeopardizing the quality of the review process.

8. <u>Facilitating Labor and Management Adjustment to Technical Change.</u> Although innovation can increase the number of workers employed within an industry over the long term, or even create an entire new industry, individual innovations may occasionally cause workers to be displaced.

In order to assure adequate time for workers and management to adjust to changes caused by innovations, I am directing the Secretaries of Labor and Commerce to work jointly with labor and management to develop a Labor/Technology Forecasting System. The System would develop advance warning of industrial changes and permit timely adjustments.

9. <u>Maintaining a Supportive Federal Climate.</u> The initiatives announced in this Message are only the first steps in our efforts to ensure American technological strength. We must also develop and maintain a climate conducive to industrial innovation. The Federal government must take the lead in creating that climate. And the Federal government's efforts must be continuing ones. I am committed to these goals.

I am charging the National Productivity Council with the continuing tasks of monitoring innovation, developing policies to encourage innovation and assisting the Departments and agencies in implementing the policies announced today. I am also establishing a Presidential award for technological innovation to make clear to this Nation's inventors and entrepreneurs that we place the highest national value on their contributions.

Each of the initiatives I have just proposed supports an important component in the innovation process. In combination, these initiatives should make a major difference in our Nation's ability to develop and pursue industrial innovation. However, these incentives will not by themselves solve our current difficulties in encouraging needed innovation. In our economic system, industrial innovation is primarily the responsibility of the private sector. The manager of the firm must decide whether to develop and market innovative new products or whether to find and employ new ways of making existing products. Although the Federal government can establish a climate that encourages innovative activity, it is the private sector that finally determines whether innovation will take place.

In addition, the steps outlined in this Message must be viewed in the context of our current severe inflation problem. With costs rising at an abnormally high rate, managers naturally have a disincentive to spend the sums needed for adequate industrial innovation. I understand and fully appreciate that changing certain of our tax laws could provide additional incentives for investment in innovation. Indeed, my approval of adjustments in the capital gains tax in the Revenue Act of 1978 has alleviated some shortages of venture capital. Many of the suggested alterations of our tax system are intertwined with other economic challenges -- such as fighting inflation. While it might be possible to make changes in the tax code that would promote innovation, these changes should not be viewed in isolation from other aspects of our economy. I will therefore evaluate tax laws affecting industrial innovation at the time that I consider my fiscal policies for Fiscal Year 1981.

CONCLUSION

Innovation is a subtle and intricate process, covering that range of events from the inspiration of the inventor to the marketing strategy of the eventual producer. Although there are many places in the chain from invention to sale where we have found modification of Federal policy to be appropriate, there is no one place where the Federal government can take action and thereby ensure that industrial innovation will be increased. We have therefore chosen a range of initiatives, each of which we believe to be helpful. In aggregate, we expect them to have a significant impact. Nonetheless, they represent only an early skirmish in what must be a continuing battle to maintain the technological strength of the American economy. I pledge myself to this task and ask the Congress to join me in meeting our common challenge.

JIMMY CARTER

THE WHITE HOUSE
October 31, 1979

APPENDIX B

The President's Industrial Innovation Initiatives, October 31, 1979

Office of the White House Press Secretary

THE WHITE HOUSE

FACT SHEET

THE PRESIDENT'S INDUSTRIAL INNOVATION INITIATIVES

BACKGROUND

The President initiated a "Domestic Policy Review" in April 1978 to identify appropriate government actions in connection with innovation. The President asked the Secretary of Commerce to lead the Review. The charge given the Commerce Department was: "What actions should the Federal government take to encourage industrial innovation?" During the course of the Review members of the Administration consulted with hundreds of groups and individuals from industry, labor, academia, and public interest organizations. Suggestions embodied in task force reports were rendered by 150 of these people. Their recommendations have been reviewed and analyzed by the President. In essence, recommendations ultimately selected by the President are designed either to develop a missing resource or influence decisionmakers in the direction of innovation.

Other industrial countries, recognizing the importance of innovation, are extending their competitive advantage through industrial policies, program, and institutional structures aimed at selected technologies. To respond to this challenge to our economy and the competitive position of U.S. industry, the review developed policy options intended to foster the Nation's competitive capability and entrepreneurial spirit for the decades ahead.

The initiatives announced today are considered by the President as first steps in meeting the Nation's commitment to innovation and the continuing challenge to maintain the technological strength of the American economy.

The President's actions provide a signal to the private sector that innovation is valued and that it is Federal policy to preserve and promote it in the years ahead. The Administration hopes this will improve the rate of innovation and will establish, over time, a climate in which it will flourish.

There are nine areas where the President has made specific decisions regarding innovation:

- Enhancing the Transfer of Technical Information
- Increasing Technical Information
- Improving the Patent System
- Clarifying the Anti-trust Policy
- Fostering the Development of Smaller Innovative Firms
- Improving Federal Procurement
- Improving Our Regulatory System
- Facilitating Labor/Management Adjustment to Innovation
- Maintaining a Supportive Attitude toward Innovation

ENHANCING THE TRANSFER OF INFORMATION

Scientific and technical information is created largely by universities, government laboratories, industrial laboratories and by similar activities abroad. It becomes the knowledge needed in industrial innovation when it is relevant to industry's problems or opportunities and when it is effectively transferred to the industry user. New actions deal with improving the transfer of existing, potentially relevant information; and improving the rate at which we create such information. To facilitate the transfer of existing information, the President is taking action in two areas.

1. The NTIS Center for Utilizing Federal Technology

The Federal government annually undertakes approximately $10 billion of R&D at Federal laboratories and Federally-funded R&D Centers. The National Technical Information Service (NTIS) provides a channel of communication with industry concerning these research results. It has a broad understanding of industry needs, and Federal laboratory activity. It is in a position to help inform industries of technological opportunities of which they might otherwise be uninformed.

- The President has decided to enhance the NTIS program by creation of a Center in NTIS with the mission of improving the flow of knowledge from Federal laboratories and R&D Centers to industries outside the mission agencies' purview. The FY 1981 cost of the program will be $1.2 million and subsequent year costs will not exceed $2 million per year.

2. Foreign Technology Utilization

Foreign technological and scientific advances are an untapped
source of technological information for American innovation.
An inadequate ability exists within the Federal government
and within industries to gather, analyze, organize, and dis-
seminate information regarding foreign research and development activities that bear on the competitiveness of U.S.
industry. Other countries gather such information on the U.S.

- The President has decided to have the NTIS include
  extensive foreign technical literature collection
  and translation in its present activities. This
  move will make relevant foreign literature available
  to industry. The first year program will be $1.8
  million.

- The President intends to have the Departments of
  State and Commerce interview volunteer returning U.S.
  overseas visitors about observed foreign technological
  developments, collect reports from our science coun-
  selors, and collect photographs, and other unpublished
  information. This information will be merged with the
  NTIS data base on foreign technical literature to make
  it widely and easily available to industry. The 1981
  cost of this program will be $2.4 million.

INCREASING TECHNICAL KNOWLEDGE

The Federal government supports a broad range of R&D activities from basic through applied research, development and
demonstration in areas of interest to industry. Most of this
work is to meet some specific social or national need, as in
the case of future development or defense, or to provide a
foundation for future advance, as in the case of basic
research. Unlike many foreign countries the U.S. does not
make major direct governmental investments in the development
of technologies. The President will take actions in three
areas aimed at enhancing the technical knowledge base in the
United States.

1. Generic Technology Centers

The President believes there is a Federal role in the development of generic technologies -- that is, technologies that
underlie industrial sectors. Examples include welding and
joining, robotics (automated assembly), corrosion prevention
and control, non-destructive testing and performance monitoring and tribology (science of lubricants). Because the benefit from advances in generic technology to any one firm (or

even one industrial sector) may be small, there is less
investment in the development of generic technologies than
would be justified by the benefits that flow from these activities.

- The President has decided to establish non-profit
  centers -- at universities or other private sector
  sites -- to develop and transfer generic technologies.
  Each of the centers will be targeted on a technology
  that is involved in the processes of several industrial sectors, and has the potential for significant
  technological upgrading. It would not supplant efforts
  in the private sector that are designed for specific
  product development.

  - Each center will be jointly financed by industry
    and government, with the government's share drop-
    ping to 20 percent or less of the center's cost in
    the fifth year.

  - Four centers will be established in FY 81 at a
    cost of $6-8 million. Three will be sponsored by
    the Department of Commerce and one by the National
    Science Foundation.

  - In future years, the size of the program will
    depend on the proposals received, and the experi-
    ence gained from this initial effort.

2. Regulatory Technology Development

One major cause of the modification of industrial processes
in recent years has been the obligation to assure compliance
with environmental, health or safety regulation. Innovation
is important in making these changes so that the new processes
meet regulatory objectives at the least cost. Federal
investment in the development of compliance technology already
is substantial. There are very large Federal expenditures on
technologies for the clean burning of coal or to improve the
safety of mines. But there are instances in which the
affected sector is unable to perform the work or to assure
speedier compliance than the market can provide.

- The President will ask the Office of Management and
  Budget, in the course of its crosscut of regulatory
  activities in developing the FY 81 budget, to examine
  closely the nature and extent of expenditures on
  compliance technology and to bolster the Federal
  effort.

3. Improved Industry-University Cooperation in R&D

The scientific and technological strength of American universities has not been harnessed effectively in promoting
industrial technological advance.   In order to achieve this
end, in FY 1978 the NSF established a program for the support
of high quality R&D projects that are proposed jointly by
industry-university research teams.

- The President has decided to provide $20 million of
  new funds at NSF in FY 1981 for this purpose with
  subsequent year support at a similar level.

- In addition, the President plans to extend the NSF
  program to other agencies. NSF will work with DOD,
  DOE, EPA, and NASA in FY 1980 and with other agencies
  in subsequent years to initiate such university-
  industry cooperative R&D programs and to establish
  quality-control procedures as effective as the NSF
  peer review system. Each agency will formulate plans
  for building its support for this program with the
  objective of reaching an aggregate of $150 million.

STRENGTHENING THE PATENT SYSTEM

Patents serve several important functions in the innovation
process. First, they provide an inventor with an incentive --
a monopoly limited in time. Second, the exclusive rights
provided by a patent can stimulate a firm to make the often
risky investment that is required to bring an invention to
market. Finally, a patent provides an important method for
disclosure of information about inventions and their uses to
the public.

1. Uniform Government Patent Policy

The Policy Review identified strong arguments that the public
should have an unrestricted right to use patents arising from
Federal sponsorship. These patents were derived from public
funds and all the public have an equitable claim to the fruits
of their tax dollars. Moreover, exclusive rights establish
a monopoly -- albeit one limited in time -- and this is an
outcome not favored in our economy.

Several competing considerations, however, urge that exclusive rights to such patents should be available. First,
government ownership with the offer of unrestricted public
use has resulted in almost no commercial application of
Federal inventions. Without exclusive rights, investors are
unwilling to take the risk of developing a Federal invention

and creating a market for it. Thus, ironically, free public
right to use patents results, in practical terms, in a denial
of the opportunity to use the invention. Second, many contractors, particularly those with strong background and
experience with patents, are unwilling to undertake work
leading to freely available patents because this would com-
promise their proprietary position. Thus, some of the most
capable performers will not undertake the government work for
which they are best suited.

As a result of the strength of these considerations, most
agencies have the authority in some circumstances to provide
exclusive rights. But because of the difficulty of balancing
the competing considerations, this issue has been unsettled
for over 30 years and the various agencies operate under different and contradictory statutory guidance. The uncertainty
and lack of uniformity in policy has itself had a negative
effect on the commercialization of technologies developed
with Federal support. As a result, there is an active
interest in the Congress and among the agencies to establish
a clear and consistent policy.

The President considered a range of options, from always
vesting title in the contractor, to maintaining the status
quo. In arriving at his decision, the President considered
the following factors:

- Uniformity. The agencies are currently governed
  either by an array of different statutes or, in
  the absence of statute, by Presidential guidance.
  Indeed, some agencies have different statutory
  guidance on patents governing different programs.
  In light of this, there is substantial confusion
  among contractors who perform R&D for several
  agencies or programs.

- Impact on Innovation. Exclusive rights to a
  patent may be necessary to ensure that a firm
  will make the often risky investment that is
  required to bring an invention into production
  and to develop a market for it. Exclusive rights
  provide protection from other firms that might
  skim the profit from the market by copying the
  invention after the risk and cost of introduction
  are reduced by the first firm's efforts.

- Administrative Burden. Any policy that requires
  an agency to make decisions imposes some administrative costs.

- <u>Uncertainty</u>. A clear and easy-to-apply rule is preferable to an ambiguous rule for the guidance it offers to both industry and government officials.

- <u>Contractor Participation in Government Programs</u>. Firms with strong proprietary positions are unwilling to accept government contracts that would result in freely available patents.

- <u>Competition</u>. Exclusive rights foreclose competition in the marketing of the invention covered by the patent and serve, in some cases, to enhance the recipient's market power.

● The President has decided to seek legislation that would establish a <u>uniform</u> government policy with exclusive licenses in the field of use. Title to the patent will be retained by the Government, but the contractor will obtain exclusive licenses in fields of use that he chooses to specify and in which he agrees to commercialize the invention. There will be an exception where the agency determines that such a license would be inconsistent with either the agency mission or the public interest. In most cases, the allocation would be after the invention has been identified, rather than at the time of contracting. The Government would license in all fields of use other than those claimed by the contractor. The Government would retain march-in rights that can be exercised in the event the licensee does not develop the patent.

● The President also supports the retention of patent ownership by small businesses and universities, the prime thrust of legislation now in the Congress, in recognition of their special place in our society.

2. Other Reforms

The achievement of the objectives of the patent system depends in large part on the strength of protection a patent provides. Today a U.S. patent has less than 50 percent chance of surviving a court challenge. Uncertainty as to the validity and continued reliability of a U.S. patent creates the threat of lengthy and expensive litigation with an uncertain outcome.

- ● To improve the presumptive validity of an issued patent, and to reduce the cost and frequency of

defending it in court, the President is proposing
several significant steps. First, the quality of
issued patents will be significantly upgraded by
major improvement of the Patent and Trademark Office's
filing and classification system. Second, he is urg-
ing the Congress again to establish a single court to
deal with patent appeals. This court would establish
nationwide uniformity in patent law, make litigation
results more predictable, and eliminate the expensive
and time-consuming forum shopping that characterizes
patent litigation. Finally, to minimize the cost and
uncertainty of litigation patent validity in the
courts, the President will submit legislation to provide for voluntary reexamination of issued patents by
the Patent and Trademark Office at the request of any
person or the court.

● One of the world's greatest stores of technical information is in the Patent and Trademark Office files,
which include more than four million U.S. Patents.
However, the current state of access to the information in these files renders their technical content
inaccessible to anyone but patent examiners. The
President is asking the Patent and Trademark Office
to undertake efforts to provide greater ease of public
access to these files. These reforms will be undertaken without an increase of public expenditures by
adjusting the fee schedule of the patent office so
that those who benefit will pay for the services they
receive. Legislation supporting these reforms will
be submitted to the Congress.

● The Administrator of the Small Business Administration
will establish an Office of Small Business Patent
Counsel to assist inventors in the transition from
invention to small business by providing the ancillary
business that attorneys rarely provide. To encourage
the development of technologically based minority
businesses, a similar office will be established in
the Office of Minority Business Enterprise and its
activities will be coordinated with the SBA. All
costs will be met by reprogramming.

CLARIFYING ANTI-TRUST POLICY

Anti-trust laws play a specific role in promoting innovation.
Vigorous enforcement of anti-trust laws spurs competition --
and the pressure of competition is a stimulant to the development of innovations that provide a competitive edge. However,

anti-trust laws are often and mistakenly understood to prevent cooperative activity, even in circumstances where it would foster innovation without harming competition.

The Domestic Policy Review revealed such misunderstanding in industry, universities, and government in instances where cooperative research is permissible, or where cooperation is not permissible.

- Industry underinvests in longer-term basic research, largely because the pay-back is difficult to achieve. In long-term research particularly, the President believes some industry cooperation is desirable. This premise led to the cooperative automotive research program, announced by the President and auto industry executives following their meeting at the White House in May 1979.

The President is taking two actions that will clarify anti-trust policy and should spur greater research activity by industry:

- The President is asking the Department of Justice to prepare a guide to clarify its position on collaboration among firms in research and development.

- The President is requesting the Attorney General, the Chairman of the Federal Trade Commission, and the Secretary of Commerce to initiate discussions with industry about innovation, anti-trust policy formulation, and enforcement. The purpose is to dispel the perception that anti-trust policy inhibits innovation and to improve communication between industry, the Justice Department and the Federal Trade Commission.

FOSTERING THE DEVELOPMENT OF SMALL INNOVATIVE FIRMS

Small, high-technology firms provide the majority of the new innovation in our economy. The major problems facing entrepreneurs in new firms have been identified as: start-up capital, second-round financing, and early management assistance. The new capital gains structure has loosened the flow of second-round venture capital from private sector sources.

In addition to other actions that generally will benefit smaller R&D firms, the President is taking four specific steps to foster innovation in small, high-technology firms:

1. <u>National Science Foundation Small Business Innovation</u>
   <u>Research Program</u>

The National Science Foundation Small Business Innovation
Research Program provides funding to small companies to permit
development of a venture analysis for new projects and demonstrate technological feasibility. The program has operated
for two years at $2.5 million. It is applauded by both the
small and big business communities. It has resulted in projects for which follow-on private-sector funding has been
pledged.

- The President has decided to expand the NSF program
  through an increase in FY 1981 of $10 million. In
  addition, the President is directing the NSF to work
  with other agencies to determine whether similar programs should be established. The Office of Management
  and Budget will coordinate development of plans and
  goals for the expansion of these programs, working
  toward a goal of approximately $150 million annual
  funding.

2. <u>Corporations for Innovation Development</u>

States or multi-state regions can join in the Federal government's efforts to spur innovation by establishing State or
regional "Corporations for Innovation Development" (CID's).
The goal is to help alleviate some of the difficulty an
entrepreneur confronts in obtaining start-up capital. These
CID's would be modeled partly after the successful National
Research and Development Corporation in Great Britain and
existing state corporations, such as the Connecticut Product
Development Corporation. Their functions would include:

- Direct equity funding for the start-up of firms
  wishing to develop and bring to market a promising,
  but high-risk, innovation.

- Guidance to potential applicants to the National
  Science Foundation Small Business Program, including serving as the second-round guarantor in
  appropriate cases.

- Early management assistance to firms that are
  funded.

- When otherwise qualified, acting as the recipient
  of Economic Development Assistance funds for the
  State or region.

- To lead the way for States or regions to establish
  CID's, the Federal government (through the Department
  of Commerce) will support two regional CID's in FY 1981.
  To provide breadth, one of these CID's will be in an
  industrial region, and the other in a less industrialized State or region. The Federal support will be in
  the form of loans of $4 million per center, on the
  condition that the region provide matching funds.

3. Federal Support for Small R&D Businesses

Funding for new R&D is a problem for small firms. The small
business community correctly believes that given their number,
and the significance of their role in the innovation process,
they receive a disproportionately low percentage of Federal
R&D dollars. To deal with this, the President is directing
each agency that contracts for R&D services to:

- Develop policies ensuring that small businesses are
  not unfairly excluded from competition for contracts.

- Publicize, through the SBA and the State or regional
  CID's, opportunities for bidding that are especially
  appropriate to small businesses.

- Report their progress toward increasing small business
  participation annually to OMB.

4. General Venture Capital Availability

As the number of new start-ups increases, the demand for
second-round financing will increase. While the capital
gains tax changes have increased the flow from taxable private
sector investors, the flow will be further encouraged by the
following actions the President is taking:

- The President is directing the Administrator of the
  Small Business Administration (SBA) to change Part
  121.302(a) of the SBA regulations to permit Small
  Business Investment Companies (SBIC's) and private
  sector venture capital firms to co-invest in a small
  firm. The changes are subject to restrictions. There
  must be an identifiable independent entrepreneur in
  control of the firm. And there must not be a provision
  for acquisition by the private sector firm as part of
  its financing.

- The Administration already has changed the Employment
  Retirement Income Security Act (ERISA) regulations to

make it permissible for fund managers to invest in small, innovative businesses. In addition, the President will request the Administrators of ERISA and the SBA to establish an interagency committee to examine what regulatory changes or other means are needed to stimulate investment in small and medium endowment funds. This will foster further availability of venture capital.

OPENING FEDERAL PROCUREMENT TO INNOVATIONS

New technology plays a critical role in promoting innovation. In a free enterprise system, however, marketplace incentives are the crucial motivators. This fact bestows a special responsibility on the Federal government, because it is the Nation's largest single purchaser of goods and services.

In the past, The Department of Defense and the National Aeronautics and Space Administration have shown convincingly the impact that Federal purchasing power can have as a marketplace stimulus. A pilot program at the Department of Commerce -- known as the Experimental Technology Incentives Program -- has demonstrated that the government can use its purchasing power to spur innovation in areas other than major systems development and high technology. The President will take actions intended to extend this experience to all Federal purchasing.

- The President is directing the Administrator for Federal Procurement Policy in the Office of Management and Budget to introduce reforms in Federal procurement practices by establishing uniform procurement policies and regulations so as to remove barriers that inhibit the government from realizing benefits of industrial innovation. Special attention is to be given to the most innovative small and minority businesses.

  - Heads of executive agencies and establishments are being asked to designate senior officials to expedite implementation of new reforms.

  - Special attention is to be given to substituting performance specifications in place of design specifications, and, wherever feasible, selection will be on the basis of costs over the life of the item, rather than merely on the initial purchase price.

- The President is asking the Administrator, General Services Administration, to expand the New Item

Introductory Schedule to publicize, within the Federal
government, the existence of new items. To accomplish
this, GSA will take steps to inform the business community -- particularly small businesses -- of the New
Item Introductory Schedule and of its benefits.

IMPROVING OUR REGULATORY SYSTEM

Government regulations often influence industrial innovation,
stimulating it in some cases and discouraging it in others.
For example, some regulations provide incentives for inventing
totally new processes to meet regulatory requirements. Other
regulations can cause industry decisionmakers to divert
resources from exploratory R&D into defensive research aimed
only at ensuring compliance with government regulations.

The Carter Administration has a record of being sensitive to
the need for a balanced approach to regulations, independently
of the Domestic Policy Review on innovation. Previous actions
the President already has taken that will have a favorable
impact on industrial innovation include:

- Deregulation of airlines and other industries.
  The President expects the pressure of competition
  to trigger innovative new ways to cut costs and
  improve service.

- In environmental, health and safety regulation,
  the Administration is emphasizing cost-impact
  analysis to take account of regulatory burdens on
  industry. The President has formed the Regulatory
  Analysis Review Group and sent to Congress last
  spring the Regulatory Reform Act to make regulations more efficient and effective.

- Last month, OMB reported substantial progress in
  the implementation of Executive Order 12044,
  which sets goals for improving Federal regulatory
  practices.

- The President created the Regulatory Council to
  provide better coordination between the regulatory
  agencies. The Council is made up of the heads of
  35 regulatory agencies. The Council is working
  to reduce inconsistencies and duplications between
  regulations, eliminate delays, reduce paperwork
  and generally keep the cost of compliance down.
  The Council publishes the Calendar of Federal
  Regulations which contains information about

major regulations under development. This is
intended to reduce uncertainty about future regulations. All of these reforms show the Administration's continuing efforts to offset negative
effects of regulation on societal objectives.

In addition to these actions already taken, the President is
announcing today several decisions specifically in connection
with improved innovation:

- The Administrator of EPA will review the agency's
  programs to determine what further opportunities
  exist to substitute performance standards for design
  or specification standards within statutory authority.
  Specification standards should only be used when they
  are clearly justified. Regulatory agencies will also
  be encouraged to explore the possibility of providing
  dual criteria for either performance and specification standards, thereby allowing individual firms to
  choose the mode best suited for them.

- In conjunction with their semiannual regulatory
  agenda, executive health, safety, and environmental
  regulatory agencies will prepare five-year forecasts
  of their priorities and concerns. Better knowledge
  of agency plans will allow industry to plan its
  research and development.

- The EPA Administrator will develop and publicize a
  clear implementation policy and set of criteria for
  the award of "innovation waivers." He will assess
  the need for further statutory authority.

- Federal executive agencies responsible for reviewing
  the safety and efficacy of products will develop and
  implement a system of priorities. Under these
  systems, the agencies will identify those products
  that are most innovative and/or have exceptional
  social benefits, and expedite their clearance reviews
  to the extent permitted by applicable statutes.
  These systems will affect the speed, but not the
  quality of the agency's review.

- To expedite the introduction of new drugs marketed
  in foreign countries and to expedite the U.S. drug
  review process, the President is asking the Administrator of the Food and Drug Administration to take
  steps to assure that our drug clearance process
  benefits from the foreign experience.

FACILITATING LABOR/MANAGEMENT ADJUSTMENT TO TECHNICAL CHANGE

Labor plays an important role in industrial innovation. Perceptions by investors of labor attitudes toward innovation influence the investors' willingness to move ahead. Labor, on the other hand, recognizes the importance of innovation and technological change, realizing that innovations that improve productivity commonly increase the number of workers employed within an industry over the long term. Labor also understands that entirely new industries have been created through innovation. Nevertheless, individual innovations often are perceived as a threat to labor because shifting skill mixes result.

The key to successful adjustment is warning time. Thus, a labor-technology forecasting system, supported cooperatively by labor and management, could be very valuable. Its purpose would be to attempt to forecast technological change within specific industries and to assess the implications for labor of such change. These forecasts and assessments could provide the basis for retraining and other adjustment activities by industry and labor. Labor has been advocating this approach for twenty years. It is long overdue. Therefore:

- The President is directing the Secretary of Labor and the Secretary of Commerce to work jointly with labor and management to develop a national Labor/ Technology Forecasting System. The President is requesting that they implement this new system in the context of ongoing labor-management activities, in conjunction with agencies responsible for adjustment assistance, and in cooperation with labor/ management teams.

MAINTAINING A SUPPORTIVE CLIMATE FOR INNOVATION

The results of the Domestic Policy Review stressed the importance of a favorable climate in the U.S. receptive to new innovation and of perceived public attitudes toward innovation. Accordingly, the President plans three actions aimed at making a clear public commitment to ensure that innovation in this country thrives in the future.

- Recognizing that future enhancements in industrial innovation lie primarily in the management/engineering area, the President is asking the Commerce Department and the National Science Foundation to host a National Conference for Deans of Business and Engineering Schools to stimulate improved curriculum development in technology management and entrepreneurship.

- The President is establishing an award for technological innovation. The existence of this award will provide explicit encouragement to U.S. industry, symbolizing a national commitment to innovation. The awards will consist of a Presidential plaque to companies in six areas: transportation, communication, health, agriculture and food, natural resources (including energy). The selection criteria will include both technical excellence and commercial impact. The Department of Commerce will be responsible for presenting the President with a list of nominees each year. The awards will be presented annually by the President's Science and Technology Advisor.

- The President is asking the Productivity Council to form a committee charged with monitoring innovation, developing policies to encourage it, assisting the agencies in implementing these policies, and pursuing the removal of legislative or administrative barriers to the innovation process.

3. Technical Capability and Industrial Competence

Two years ago the Swedish Royal Academy of Engineering Sciences was asked by the Government to make a study of the status of Swedish technologies, the challenges that face our industry now and in the future, and the competence that is needed to meet these challenges. This is the largest study our Academy has ever performed. It was completed in the summer of 1979, and the results were presented in five volumes. Two hundred fifty Academy members and 750 other persons took part in this study.

Introduction

Sweden's industrial production has increased a hundred-fold during the last 100 years, which corresponds to an annual growth of more than 4%. Since 1945, a fivefold increase in volume has occurred. The main factors behind this development have been an ample supply of domestic raw materials and hydro-electric power, the high technological level of Swedish industry, and a dynamic business leadership. These factors have been effectively exploited through active participation in international specialization, which has resulted in a rapid build-up, such as flexibility and adaptability, many unique inventions, and the fact that Sweden was not involved in either world war.

Swedish industry and commerce presently face a strong challenge after several years of crisis. The crisis has demonstrated long-term changes in Sweden's development potential and competitiveness. In particular, these changes have been brought about during the post-war years, principally by

- the substantially decreased contribution to exports and national product from raw materials and semi-finished products, and

- the loss of the technological lead that important
  sectors of Swedish industry reached at the end of the
  second world war.

There are coincidentally many structural problems in
Sweden which will require strong, effective, and goal-oriented
action from both industry and government. With this background the Royal Swedish Academy of Engineering Sciences
(Ingenjorsvetenskapsakademinen – IVA) was commissioned to carry
out a study of Sweden's technical and industrial competence.

The basic issues of this study can be summarized with the
following questions:

- In what respect have long-term changes in the conditions for the ability of Swedish industry to expand
  and compete been brought about during the post-war
  years?

- What kinds of adaptation do these changes require in
  structure and performance of Swedish industry?

- What is, in this connection, the role of technical
  and industrial competence? Is further strengthening
  and broadening required?

To execute the study, the Academy appointed a Main
Committee, which in turn had four Special Committees, to deal
with specific areas, namely:

    I.   Industrial development during the post-war period –
       the role of technology,

   II.   Technical and industrial competence – international
       comparison,

 III.   World trends in technological development, and

  IV.   The potential for development of Swedish industry in
       an international context.

Each of the Special Committees prepared reports and other
background documents for the Main Committee's consideration.

This summary is based on the Main Committee's final
report. The final report's introduction describes the background and presents the set of problems which initiated the
study. In this part, some consideration is given to the
relations between employment and technological development.

The second part is a basic analysis of the long-term problems
of technical/industrial competence and future competitiveness
of industry. It also describes some of the threats and possibilities arising from trends in international trade and technical development. The third and final section presents the
Main Committee's conclusions and recommendations.

The Historical Development of Swedish Industry

Basic to this study is an analysis of the historical
development of Swedish industry and society through the
country's industrialization process. Five periods are dis-
tinguishable:

- up to 1890 - this period is typified by dominance of
  agriculture, the rapid growth of saw milling and local
  industry based on metal working;

- 1890-1913 - this period is characterized by increased
  product refinement and the foundation of present-day
  Swedish industry;

- the inter-war period is best described by the con-
  solidation of industry, and Sweden became a net
  exporter of capital;

- 1946-1965 - the "golden industrial age" in which productivity increased as at no other time; however, at
  the same time profitability and capacity for self-
  financing decreased; and, finally,

- 1965 to the present.

The post-1965 period is of course particularly important
because of its strong internationalization and deteriorating
growth conditions. In 1965-1970 industrial growth rates were
annually only 5%, and in 1970-1975 this rate declined to only
2.4%. As employment in industry fell, and the weak international situation began to affect exports, Sweden's trade
balance deteriorated. Investment was diverted to industrial
rationalization and capital needs intensified as at no other
time. From 1970 to 1975 the same trends continued, but the
growth rate of overall productivity decreased by 50 percent.
For the first time since Sweden became industrialized, the
input of resources fell.

The current crisis has similarities with previous crises
in the 1920's and 1950's, but the differences are dramatic.
Generally, structural problems are more severe. The oil price

situation caused the shipbuilding market to weaken even more, and this was compounded by blows to the steel industry. International investments were reduced, and there were few apparent growth industries to take up the slack. Furthermore, the self-financing capability of industry sunk to 20% -- mainly because of low profitability. Various safety nets of subsidies, credit guarantees, and "soft" loans reduced the ability of industry to quickly adapt to the changing conditions. This combination of events impacted Sweden more severely than it did to most of her competitors.

Sweden's contributions to international technical development are modest on an absolute basis. However, on a comparative scale which considers Sweden's small population, her advances have been considerable. Further, in certain industrial sectors, Swedish technology has been outstanding. Examples are to be found in high-voltage direct-current transmission, high pressure processes, thyristor-controlled locomotives, enrichment and transport of ferrous minerals, electro-steels, and ship design.

The new competitive situation facing Swedish industry is attributed to the following key factors:

- reduced marine freight rates which make raw materials less expensive than in other parts of the world;

- the availability of pulp and paper from the U.S. and Canada which increases their share of the West European market;

- the development of new, cheaper, and high quality iron ore deposits in other countries such as Brazil and Australia;

- the emergence of Japan and the newly industrializing countries (NICs) which increasingly compete against Sweden's classical products;

- the phenomena of the increasingly rapid transfer of technology;

- shorter technological lives of products; and,

- requirements for greatly increased return on invested capital.

In addition, the development of large multinational

companies and banks has facilitated the transfer of management know-how and capital, thus increasing competition.

The Status of Swedish Education and Research

Technical competence depends heavily on education;
therefore, the report looks at education trends in secondary
schools. It has been suggested there has been some reduction in standards in science teaching. As a result of
broadened admissions, more students are entering universities,
but the failure rate is higher than previously. Also, more
students are choosing a less comprehensive curricula, and en-
rollment in science subjects at non-technical universities
has fallen considerably. Fewer students are studying engineering than ever before. To compound the problems, the average
time taken to achieve the first degree is increasing.

The study has examined the role and meaning of research
in the Swedish society and in similar OECD countries. In 1977
the total cost of research and development in Sweden was Skr
7 200 million at current prices. Sweden is characterized by
a "sectoral" science policy, that is to say, there is no central
government research and development budget, rather each governmental department commissions its own applied research and
development from its own budget and according to its own
priorities. The report recognizes at the outset that central
government has, however, a responsibility to support basic
research.

Research at universities and technical institutions has
been examined critically. Basic financial support for universities is provided by the Ministry of Education. The development of these grant programs over the last few years has been
examined. The analysis shows that the level of funding for
basic finance to technical universities has fallen substantially. This has created difficulties in maintaining the level
and standard of university research, as well as supplying the
growing need for competent staff in sectoral research.

The linkages between universities and industry and the
possibilities for contract research being carried out for
industry in universities and research associations are considered in the report. The report points out that only 1%
of the world's research and development is carried out in
Sweden. It also stresses the crucial need for Sweden to
keep in close contact with overseas research and development
sources. The need for research and development planning at
central level is highlighted. Sweden is very similar to other
countries in this respect.

Swedish Industry Today

The report then turns to an analysis of the Swedish
industries' current capabilities and development possibilities
beginning with a look at the present situation. This is
typified by Sweden's relatively large number of international
firms with leading positions in specific market sectors which
command a noteworthy large export ratio of 40%. About 80% of
Swedish exports go to other industrialized countries in
Western Europe and North America, and this involvement with the
other industrialized countries is inextricable. Sweden's
investment in R&D is relatively more than her competitors,
yet she lags in markets served by them.

Industrial strategy in Swedish industry was studied by
means of interviews held with twelve large companies. Generally, these firms had altered their strategy in the 1970's
to cope with reduced utilization of installed capacity and
reduced profitability. The previous trend to diversification
had virtually stopped. Important features of the new strategies include: to become leaders in their specialized technological areas; to define new areas of application for their
specialties; to develop specific products into systems; and,
to increase financial strength with a view to international
creditworthiness.

New developments in Swedish industry show several interesting features including: new products and production
methods, flexible production techniques, the development of
volume markets and exploitation of market niches, and the use
of targeted research and development. The increased element
of risk taking, the need for an adaptable company organization,
and the role of long-term planning are discussed in the report.
Attention is given to technology transfer and the specific
problem of founding small technology-based companies in Sweden.
The unavailability of risk capital; the mass of new legislation
and the complex governmental bureaucracy are cited as factors
which cause difficulties and are discouraging to the formation
of new, small businesses.

Some particular problems in industry are highlighted. In
the raw materials (primary) industry, there are major problems.
However, it is noted that the technical competence level is
high and this should be utilized to improve manufacturing processes in order for this industrial sector to become more competitive internationally.

With respect to the manufacturing (secondary) industry
sector, it is pointed out that criteria for selection of new

fields in which to compete must be carefully chosen. Criteria
for selection must include: an expanding market, a knowledge
intensive sector, and a field in which the company can establish a well-known name. Furthermore, the companies should
preferably deal with products insensitive to price changes.
Swedish industry problems in broadening its market base include: the demands for local manufacture and direct investment
overseas, the heavy manpower demands entailed in exporting
"systems," and the difficulty of financing large export
"packages."

A generalized market strategy is suggested in the report.
Markets in the technically advanced countries should first be
attacked with technically advanced new products. Markets
should preferably be those in which Swedish industry has already made investments or is about to invest. The competition
from newly industrializing countries should not be met by
competing directly in their chosen areas. The adaptation of
new products for foreign markets is essential, especially with
respect to standards and local demands.

Innovation and the role of government assistance is the
subject of special inquiry in the study. Historical Swedish
factors and international comparisons are discussed in the
report. Technology transfer is considered a factor in developing an industrial policy. The role of government procurement can be crucial to such a policy and is examined.
The need for a collaborative development effort between the
manufacturer and the customer in this respect is discussed.
Furthermore, the importance of general support measures, as
opposed to selective ones, is stressed.

The biggest impediment for establishing new technology-
based small firms in Sweden is the lack of venture capital.
There is also a shortage of entrepreneurs because incentives
to establish such firms are not very strong. The problem is
compounded because the general public's attitude towards
industry is not too favorable at present.

Government support to small and medium-sized firms is
mainly based on regional policy. Responsible agencies lack
experience in the special problems concerning development of
new business in high-technology areas. Several measures are
suggested in the report to provide new technology-based firms
with venture capital and management support.

Trends in International Trade and Industry

The trends in international trade and industry were
analyzed in the study, and some major factors were identified:

the new ease of spreading technology, a growing trend to
protectionism, and the increasing debt of the developing
countries. Two major components of technology diffusion are
commercially based deals, which may involve finance and the
import of capital goods, and by the general increase of
knowledge through education. It is pointed out that Sweden's
experience of commercial technology transfer is limited.

"Non-tariff barriers to trade" and other forms of protectionism are increasing in spite of the reforms by GATT
which will hopefully lead to improvements. The demand for
local manufacture as opposed to efforts is seen as a factor
here. Easy access to export credits and increased international liquidity have made competition tougher for Sweden.

The outlook for industrial development and for rapid
economic growth in Sweden is not good. Several long-term
factors may be identified:

- a stable population level with a stable number of
  economically active people. Mobility between economic
  sectors is low, and the population is placing increased value on leisure time due to the effects of
  the fiscal system,

- a trend away from free trade to restrictive trading
  patterns,

- rapidly increasing energy costs will have effects on
  metals and petrochemicals production and lead to re-
  substitution of manmade materials by natural products,

- a need to safeguard natural resources and raw materials will lead to better designed, longer-life products and also to increased recycling,

- changing social demand patterns, specifically for
  services such as education and health, and

- the vastly increased sector concerned with gathering
  and transferring information for the private and
  public sectors.

New Competitors and Markets for Sweden

Developing countries were looked at from two points of
view, as competitors to Sweden and as markets for Swedish
exports. Both technical (know-how) and financial factors were
considered.

In countries that focused early on industrial development (Israel, Singapore, Greece, Hong Kong, Portugal, Taiwan,
South Korea), the importance of multinational companies is
discussed. The tendency of those countries to produce low-
priced mass-market products is recognized. These countries
are now moving into more advanced technologies, such as
communications equipment, electronic and electrical goods,
and medical equipment. Most of the growth is attributed to
imported know-how. Their consumer demand pattern is similar
to the industrialized countries. Housing and health care,
however, remain "interesting" undeveloped markets.

The larger semi-industrialized countries (Spain, Yugo-
slavia, Argentina, Brazil, Mexico, and Turkey) are expected
to reinforce their efforts to export capital goods, chemicals,
and other intermediate products. It is considered that these
countries will be competitive in steel, petrochemicals, and
ships. Typical infrastructure investments which provide
marketing opportunities are transport and traffic equipment,
communications, and energy distribution systems. Regional
development schemes involving exploitation of resources are
important. The Swedish construction industry has potential
opportunities in these markets, but it must develop the
markets more aggressively.

The development of countries rich in natural resources
(e.g. Venezuela, Iran, Chile, Malaysia, Tunisia, and Thailand)
is to some extent held back by fluctuations in commodity
prices. They need technology to develop their resources such
as oil, minerals, and timber. They also need a better developed infrastructure as well as consumer goods.

The petroleum exporting countries which have surplus
capital are similar to those rich in other natural resources.
They are, however, financially better off, and this is reflected in their purchases of sophisticated and automated
materials and equipment. For example, the development of
petrochemical industries in those countries will have a
significant impact on the chemical industry of industrialized
countries.

The final grouping is of large, poverty-stricken countries.
These countries depend on bilateral and multilateral aid, and
Sweden must work to develop goodwill in those countries which
will lead to long-term market growth opportunities. To a great
extent, the development of markets relates to "basic needs"
of which food, housing, and health are key components. Small-
scale decentralized technologies are also important to those
countries.

World Trends in Technological Development

In the analysis of the future prospects of Swedish industry, worldwide technological trends and their consequences
for Sweden were assessed. Sweden's relative position had to
be defined regarding acquisition and/or development of new
technologies. This was necessary both to provide a background for the Main Committee's recommendations for governmental measures in research and education and to indicate
directions for further industrial efforts in promising
areas.

The whole membership of the Academy was engaged in this
part of the study. Thereby access was gained, unique for
Sweden, to technological competence experienced in industrial
management, production, and marketing. About 250 experts thus
participated in an enquiry into the development potential in
about 100 technical fields, including energy, electronics,
transportation, materials, chemical industry, biotechnology,
health and environment, agriculture and forestry, production
technology, and construction.

For each of the 100 potential developments, possible
trends were described, and the following questions were
asked:

●   Is the development probable, and if so, in what time-
    frame will there be an industrial break-through?

●   Does Swedish industry have opportunities in the field;
    what threats does it involve?

●   What is the overall competence in the field in Sweden
    today?

●   Should Sweden allocate special development resources
    to the field?

Three dimensions are used to describe the scope and
importance of Sweden's action in each of the areas; namely,

●   the degree of importance for Sweden to engage
    development resources in the field of interest,

●   the type of efforts needed by industry and by the
    research community, and

●   the time scale in research and the necessary degree
    of short-term industrial preparedness.

The findings are summarized according to these dimensions. In the findings strong emphasis is, for example, given to

- basic knowledge in materials sciences,

- application of new production technology and utilization of new construction materials,

- equipment for telecommunications and process control based on modern microelectronics,

- biotechnology,

- technology for health care, and

- technology for resource saving and environmentally advantageous production processes in all branches, especially in the energy sector.

Conclusions and Proposals

Sweden's Adaptation to Change-Improving Industrial Performance

The final section of the report – the Main Committee's conclusions and recommendations – opens with a discussion of Sweden's adaptation to changes in the outside world. The study adduces two problems, the first looks at Sweden's possibilities to outweigh the long-term negative changes in traditionally important areas through increased efforts in other fields. The second issue is that of society's flexibility and adaptability. The post-war years' increased dynamism abroad has reinforced the need for adaptation. At the same time, however, adaptive mechanisms have become much less effective in Sweden, especially since 1965.

To face the outside world and the future, Swedish industrial business must develop further. Expansion and competitiveness, primarily in the short and medium term, must be attained within the already existing industrial structure. Only in the longer perspective can entirely new fields of activity become important to production and exports.

The view is that Sweden's success in the foreseeable future will depend on existing large and medium firms' (a) developing products, (b) broadening their market, and (c) increasing productivity in their existing areas.

The only industrial sector in Sweden to have experienced uninterrupted growth since 1945 is fabricated metal products, machinery, and equipment. Several measures are needed to

maintain growth, including: meeting competition with improved
competence; developing "niche" strategies, establishing a
high degree of standardization, and searching actively for new
markets in the rapidly expanding middle-income countries.
Governmental innovation policies may be required in some of
these respects. Research and development is an important
strategic factor.

The raw materials industry will continue, for several
years to come, to account for a considerable share of Sweden's
industrial development. Resource-based industry still has
comparative advantages because of its geographical closeness
to major markets, the quality of its resources, and its high
technical competence. For the minerals industry, some very
specific recommendations include: development of effective
techniques for processing aluminum and uranium shales; increased prospecting for the alloy minerals such as nickel,
cobalt, vanadium, and tungsten; provision of more resources
to increase geological and prospecting capability; and the
introduction of a specialized university degree in geological
technology.

The potential for growth and competition must be skill-
fully developed, particularly in the forest and mining industries. Emphasis must be placed on increased and improved
education.

A large, viable sector of small and medium-sized businesses and the active development of new firms are essential
factors to increase flexibility as well as to broaden business
life in general. Intimately connected with this is an overall
improvement of the innovation climate, which calls for governmental support measures.

The contribution of an even more qualified and involved
workforce in manufacture is noted - not least through the
possibilities of decentralized production units, which give
an individual worker a more positive input. Sweden's experience with this feature gives her an advantage over her competitors.

In exploiting new markets which have different trading
methods and structures, it is important to develop new forms
of cooperation between the Swedish government and industry.
It is also important to engage in new joint ventures, particularly in newly industrializing countries. Given a suit-
able long-term strategy, these markets will present more
opportunities than threats. Selling in developing-country
markets frequently requires government assistance, and it is
strongly recommended that Sweden's surveillance of these

countries be improved through better technical and economic competence at embassies; more expert missions and
seminars; and a widened exchange of scientists, technical
experts, and technicians.

Exports of systems, both hardware and software, have
become increasingly important in world trade, but still make
up only 8% of total Swedish exports. Here, Swedish firms
are called upon to increase their efforts. Where technical
competence in the buying country is insufficient, attention
should be given to modular systems to meet buyers' needs and
ability to utilize.

Strengthening the Knowledge Base

Knowledge is considered the fundamental basis for future
competitiveness.

For schools, it is recommended

- that quality be improved through wider choice of
  courses, better preparation in mathematics and languages, and alteration of the study selection system
  to encourage more pupils to enter the natural
  sciences branch,

- that technical experience and training of teachers
  be improved by way of greater contact between schools
  and industry and that efforts be made to increase the
  interest of teachers in technology and natural sciences,

- that experimental sciences be introduced earlier in
  the curriculum, and

- that more females be encouraged to study technical
  subjects and to go into industry.

For education at technical universities, the Committee
makes the following recommendations:

- New main and complementary courses should be introduced to increase students' competence in administration, economics, languages, biology, social organization, and legal matters.

- Internationalization of technical education should
  continue and be reinforced, for example, by widening
  exchanges between Swedish and foreign technical
  universities.

- Industry- and technology-oriented public and private bodies should be given the opportunity to influence the planning of higher technical education, thus complementing views put forward by the Ministry of Education.

For higher and continuing education in general, the following recommendations are made:

- Opportunities for complementary education should be improved, and financial support to students for higher degrees should be increased.

- The nominal time for securing the doctoral degree in technology should be reduced to 3 years to obtain better compatibility with the requirements of industry.

- As industry's competitiveness is enhanced by higher education, companies as well as public bodies should recruit more people with higher degrees.

Technical and scientific research is a key factor in maintaining and increasing competence and should, in the Committee's opinion,

- Considerably more resources should be made available for R&D.

- Competence must be built up in technical areas judged important for the future even though there may not yet exist a Swedish industrial base in those areas.

- The employment conditions, including administrative load, for researchers should be improved and simplified.

To help industry's needs for research collaboration and information the following suggestions are made:

- The development of an effective system for contract university research should be considered further.

- The number of part-time industrial professors should be increased, and opportunities should be created for teachers/lecturers to work in industry.

- Restrictions and disincentives, including fiscal ones, to international research and development should be removed and increased funding be made available for foreign travel and study visits.

- Possibilities for organizing centers of excellence in industrial or branch research should be considered.

- Considerably more resources should be made available for providing information and documentation.

New Technological Prospects

This and other studies by the Academy show that Sweden has been able to maintain her traditional strength in established technological areas, even though other countries are catching up. It has also been observed, however, that Sweden has a considerably lower competence than her leading competitors in a number of key areas. Therefore, a thorough analysis has been made of measures that can be taken to keep up and improve Sweden's competitive situation. Further efforts in different scientific fields as well as new industrial activities will be necessary. A brief list of important technological fields for potential future development follows.

- Energy

- Electronics, communication and information processing

- Materials sciences

- Chemical sciences and chemical industry technologies

- Biotechnology

- Environmental technology

- Technology for agriculture, forestry, food, and forest industry

- Production technology

- Technology for health care

- Construction and building technology

Governmental Support and Control Measures

Our report recognizes that in order to secure the future competitiveness of Sweden, private and public sectors will need to work together. As a fundamental step, it is considered essential to develop an industrial strategy and policy. An element of any such policy is support of innovation.

Government plays an important part in the support of development of new technology, including encouragement of innovation. The study recognizes, however, that the choice of technology and business areas lies with the companies. Large governmental projects to boost new technological areas are not favored. It is recommended that general measures of support include the specific tax deduction allowance for research and expenditures be raised from 10% to 20%, and that company-related R&D investment funds be introduced.

Further, government participation is of crucial value in developing certain types of large projects. Measures called for include

- considerably increased support to boost technology procurement, involving public bodies and private industry. This support, it is suggested, could come from the fund for aiding large projects recently proposed by the Government.

- a strong effort on the part of central as well as local governments to increase their competence in and the use of technology.

Specific measures suggested to help development of technology-based smaller firms include:

- Commercial banks should be given an extended role in exploitation and development activities, possibly by changes in bank legislation.

- A new Development Fund jointly organized by the state and banks should be introduced, especially for supporting the establishment of inventive small businesses.

- Support should be given for the establishment of joint public/private development companies, which would provide assistance to technology-based small firms. The Small Business Investment Corporation in the United States is an example.

Long-term Studies and Research and Development Planning

In order to assist the long-term structural development of Swedish industry, broad judgments and assessments will be needed. Up to now, there have been either sectoral studies, such as on long-distance transport costs and changes in the international capital market or very specialized company studies on specific products or markets. The argument is put forward for coordinating and consolidating the numerous Swedish studies and for taking account of prognoses made by international organizations such as OECD and the Council of Europe. Technical development trends, it is suggested, should be periodically reviewed and critically judged using refined methodologies. Areas of need/demand, complement studies of technical development, and such demand patterns should be studied and analyzed. The quality of such studies should be raised as high as possible. The Committee recommends that, in order to coordinate and finance these studies, a special fund should be established by an institution with close connections to the government. The board which oversees this fund should have representatives from trade and industry.

These studies should be utilized by the government in making overall judgments of the long-term development of Swedish industry. Such studies are utilized by the French government in its planning process.

For national research and development planning, it is suggested that:

- Long-term R&D planning should be developed in the direction now indicated by the government, including the introduction of three-year R&D programs.

- The responsibility for national planning of R&D should lie with the Prime Minister and a specially appointed permanent secretary in the Prime Minister's Advisory Committee.

- A new Research Advisory Committee should be organized and should include the Prime Minister, the Ministers of Industry and Education, representatives of research institutions, industry and commerce, and other users of the research.

4. The Committee for Economic Development Report on United States Technology Policy

I am pleased to join in this endeavor to bring together the essence of a number of intensive studies of technology and industrial innovation. The very creation of so many broad studies of this subject so close together in time is in itself significant. It tells us, I believe, that the initiators of these studies have felt strongly that technological innovation can do much to improve life, but that its actual contribution is being held far below its potential, and at least partly by factors we could change. This certainly describes the view of those of us who have participated in a recent Committee for Economic Development (CED) study of the subject.

Background of the CED Study of Technology

The Committee for Economic Development is a national, nonprofit, nonpartisan research and educational organization formed in 1942, whose trustees are mostly top corporate executives and university presidents. CED has long been concerned with the development of public policies to help bring about steady economic growth at high employment and reasonably stable prices, increase productivity and living standards, and improve the quality of life and opportunities for all Amercians.

During this century, technological progress has played a crucial role in achieving our nation's enviable record of economic growth. New products and processes and better and more efficient methods of production have turned the promise of a better life into a reality for generations of Americans. During the last decade, however, serious signs have emerged indicating that U.S. preeminence in technology and innovation is being threatened. Other nations, particularly Japan and West Germany, are diminishing the U.S. lead. Capital investment as a percentage of output is now substantially lower in the United States than in our major industrial competitors.

In the United States itself, certain public policies actually discourage new capital investment and have caused business to shift some of its research and development efforts from longer term goals to more defensive, short-term strategies.

Unhappily, inflation, low productivity, trade deficits, and job loss are all linked to inadequate technological growth and innovation. Several years ago CED became concerned about the potentially severe impact of an innovation decline on broader economic objectives, particularly on efforts to slow inflation and improve productivity. CED set out to analyze what could be done to reverse the decline.

The study which resulted, <u>Stimulating Technological Progress</u>, is, I believe, a substantial contribution to the debate now ongoing within the Carter Administration, on Capitol Hill, and in the various research communities (1). What makes this study unique, I believe, is its knowledgeably weighted judgments as to the relative importance of techno- logical innovation and the relative priority among actions that might be taken to improve it.

To develop these judgments, CED brought together a talented group, the voting members, which consisted mainly of senior executives of high technology firms and heads of major research universities. This CED subcommittee was aided by a number of experts from universities and research centers who participated actively in their deliberations. I believe the conclusions the group has reached will attract support from a widening circle of thoughtful Americans as the months progress. A list of the names of all persons in the CED study appears in Appendix A.

For input the CED group reviewed a large number of factual studies of various aspects of technological develop- ment. Some of the most insightful evidence, however, came in the form of the knowledge of our participants concerning the factors influencing decisions within individual organi- zations as to whether or not to undertake research, and whether and how to apply its results. I shall not take the space to do more than exemplify what we regarded as the relevant evidence. For more detailed support I refer you to the full policy statement.

Let me now outline what I regard as the key judgments that the group reached.

First, and most generally, we concluded that stimulating technological progress does indeed deserve to be in the top rank of U.S. economic priorities. To be sure, promoting

innovation is not always mentioned in the same breath as controlling inflation, maintaining a high level of employment,
achieving a lasting solution to our energy problem, or improving real living standards both here and abroad. Surely these
four are among the most important economic issues facing the
nation as we enter the 1980's. The successful solution of
these four dominant issues will involve changes in many
policies -- among them macroeconomic monetary and fiscal
policies and also many microeconomic government interventions
in various sectors of the economy.

During the last two decades, CED has studied these basic
economic problems and proposed policy changes to help solve
them (2). Our studies and observations leave us deeply concerned, however, that most policymakers and the constituents
they try to satisfy are seriously underestimating the potentially significant role that technological innovation can
play in achieving long-run solution to these major economic
problems in the United States

The Contribution of Greater
Technological Innovation

Because underappreciation of innovation saps public
interest in encouraging it, let me amplify this point. As
CED sees it, technological innovation is instrumental to the
long-run solution of a number of our major economic problems
including:

- Improving Real Living Standards. Technological change
  is judged to have been the primary source of rising
  living standards in the United States since 1900.
  Many studies attribute between one-third and one-half
  of the growth of real per capita income to technological change (3). Only the improved quality of the
  U.S. labor force -- a development that is itself
  partly the result of learning new technology in
  schools and colleges -- is given credit for a larger
  part of economic growth. Capital accumulation
  usually appears as the third major contributing
  factor, although its influence is often associated
  with technological change and vice versa. In fact,
  there is a significant overlap between greater application of capital to productive processes and technological change. Greater investment in plant and
  equipment usually incorporates the knowledge of improved design in equipment and techniques.

  There are, of course, adverse side effects from some
  innovations, sometimes promptly perceived and some-

times long deferred. Great industrial and technical
strides have sometimes caused pollution, new health
problems, job dislocation, and social disruption.
When we have become wise enough to be properly concerned about them, however, other technological know-
how has often proven to be the most effective way to
remove or curtail the bad effects. In this arena as
in many others, it is important to judge relative
benefits and net gains. In the past several decades
the overall quality of living in all industrial
countries has been substantially improved through
greater use of technology. Technology has been and
can be employed to abate pollution, conserve natural
resources, reduce working hours, enhance education,
improve the quality of health care, expand the worth
of family assets, and bring artistic performance to
large numbers of people. Technology, if properly
directed, can help society deal with its problems
and can increase benefit-cost ratios in the process.

● Helping to Control Inflation Through Increasing
  Productivity. Insufficient productivity growth is
  now widely recognized as a serious problem in the
  U.S. economy. Since 1973 there has been a significant
  decline in productivity growth rates in most economic
  sectors. For example, manufacturing productivity,
  which was improving at an annual rate of 2.4 percent
  during 1965-1973, slowed to a growth rate of only 1.7
  percent for 1973-1978. The adverse trend in other
  industries, such as mining and services, has been
  even more pronounced; in fact, in mining the measured
  level of productivity has actually declined. Meanwhile, attempts to keep wage levels increasing faster
  than this sluggish rate of productivity growth has
  served to increase inflationary pressure.

  In general, inflation occurs when demand for goods
  and services outstrips supply. Given time, technological advance can reduce long-lived inflationary
  pressures by raising productivity and thereby increasing the output produced by given resources. While
  the fruits of technological advance cannot normally
  be realized quickly enough to offset short-run demand
  excesses, its gains cumulate impressively over time.

  In earlier decades, fears were voiced that rapid
  technological change would lead to large-scale technological unemployment. But these dire predictions
  have not been borne out, and such fears have subsided.

In fact, studies show that employment tends to grow
more rapidly in and around more technologically
progressive industries because innovative activity
involves additional expenditure for capital and labor.
In addition, commercialization of the innovation
eventually leads to decreased costs that permit increased production and therefore increased employment
(4).

● Contributing to a Favorable Balance of Trade and
Protecting U.S. Jobs from Foreign Competition. For
the same reason that technological progress permits
expanded domestic consumption, it also permits expanded foreign consumption and therefore expanded
exports. Technological change encourages new or
improved products and keeps prices down, thus making
U.S. goods more competitive with those of foreign
producers. Government data indicate that since the
1960's, the U.S. balance of trade has been better in
technology-intensive products than in other products
(5).

One might hypothesize that U.S. job gains could be
even greater if our innovations proceeded unhindered
while advanced technological knowledge was withheld
from foreign competitors. But that ought not to be
thought of as a viable option in an increasingly
interdependent world with multiple sources of technological know-how. Furthermore, its consequences
would ultimately slow growth in world living standards
everywhere, including in the United States.

● Shifting to More Dependable Sources of Energy.
Probably in no problem area is our dependence on
technological advance for an enduring solution being
more dramatically displayed than in energy. To
achieve a gradual reduction of dependence on Mideast
oil, we look for technological applications to accelerate domestic fossil fuel production and hold down
the accompanying adverse environmental effects. And
for the long-range transition to major reliance on
renewable (or virtually inexhaustible) sources of
energy, we wait upon technological break-throughs to
bring us abundantly unable solar power and energy
from "clean" nuclear fusion.

The CED policy statement does not claim that technological progress is the "silver bullet" which will
solve all of our economic problems. But any policy

strategy which does not put a high priority on greater
technological innovation will fail, in our judgment,
to achieve the structural changes necessary to
effectively deal with inflation and sagging real
income. In addition, unless innovation is stimulated,
we do not believe that our civilization will be able
to meet the pressing social demands that it is
generating.

Identifying the Innovation Problem

In the face of this evidence concerning the value of
technological innovation, we find it incongruous that U.S.
investment in innovation is less than it could beneficially
be. But that, in our judgment, is the case.

Our innovation shortfall does not seem to stem equally
from all five phases of the innovation process: namely, basic
research; applied research; development of commercial feasibility; commercial introduction of the innovation; and its
diffusion through the rest of the productive system.

All five phases of the innovation process are important,
but they are not equally nurtured. The United States, in fact,
continues to make a fairly significant investment in research
and development (R&D). This year well over $50 billion will
probably be spent on this phase of the innovation process.
Compared to our major industrial competitors, we are spending
roughly the same proportion of our gross national product on
total (defense and non-defense) R&D. However, despite recent
increases in enterprise-funded R&D, we invest proportionately
slightly less in non-defense R&D than Germany and Japan.
Moreover, there are signs that more of our R&D outlays are
financing what might be termed "defensive research" -- research to protect existing investments, to respond to regulatory requirements, or adapt to much higher-cost energy.

Without denying the importance of R&D and the need for
some improvements in this policy area, the analysis underlying
the CED view of the problem clearly suggests that the greater
part of our problem of inadequate technological progress lies
in the later phases of the innovation process. Lack of investment in innovative plant and equipment has been an especially serious problem for the United States.

- For almost twenty years U.S. new investment in
  private physical capital other than housing has been
  running proportionally less than in all other major
  industrial countries including the United Kingdom.

- Such capital investment as a proportion of output
for the total economy and for manufacturing has been
substantially below our major industrial competitors.
For example, since 1960 the United States has invested
9.1 percent of manufacturing output in new manufacturing plant and equipment, compared to 15.9 percent for
Germany and 28.8 percent for Japan. (See Table 1.)

All industrial nations have relied heavily on applying
more capital to productive processes to achieve economic
growth and raise productivity. The disturbing fact is that
in the United States since the early 1970's there has been a
slowing trend in the growth of the capital/labor ratio.

We are all aware of the difficult technical arguments
over whether this slowing is purely a falling off in "capital
deepening" or whether it is "a decline in innovation." However, from the magnitude of the difference between the slow
rate of capital investment in the United States as opposed to
other countries, it is clear that lack of capital formation
is the major part of the innovation problem. No matter how
many resources we devote to R&D, there will be no technological progress unless there is investment in new plant and/or
equipment which embody the fruits of that R&D. In addition,
there is a substantial body of research which suggests that
a high level of capital investment is one of the strongest
stimulants of research and development.

The innovation problem is, of course, compounded by the
recent dismal performance of the U.S. economy, particularly
on the inflation front. Inflation averaged a mere 2.1 percent
annually in the 1958-1968 period. During the 1970's, inflation tripled to a 6.5 percent annual rate and climbed into the
double-digit range in 1974 and 1979. A sluggish economy
sparked an increase in unemployment from 3.5 percent in 1968
to 6.0 percent in 1978. The annual growth in private sector
labor productivity fell from a 3 percent level during 1948-
1965 to 2.0 percent during 1966-1973, and since 1973 it has
remained below 1.0 percent. As a result, real income grew
slowly during the 1970's, and real GNP grew only about 2.8
percent annually, compared to 4.5 percent a year in the previous decade.

Business investment, which is crucial to continued technological progress, rose only 2.6 percent a year in real terms
from 1968-78, compared with a 6.2 percent annual average in
the preceding ten years. The real rate of return on investment has declined dramatically since the mid-1960's -- from
a 10 percent level in 1965 to the current 5 percent.

Table 1. Capital Investment, Excluding Residential Construction, as a Percent of Output, Six Countries, 1960-1977.[1]

Item and Years	United States	Canada	Japan	France	Germany	United Kingdom
Total Economy						
1960-69	14.9	20.0	28.8	19.5	20.1	16.5
1970-77	14.5	19.3	26.7	18.8	18.7	17.6
1960-77	14.7	19.7	27.8	19.2	19.5	17.0
Manufacturing						
1960-69[2]	8.8	14.4	29.9	NA	16.3	13.4
1970-77[2]	9.6	15.1	26.5	NA	15.2	13.6
1960-77[2]	9.1	14.7	28.8	NA	15.9	13.5

NA = Not available.

[1] Fixed investment at market prices as a percent of output at factor cost. Current prices.

[2] Period ending 1974 for Japan and 1976 for Germany.

Prepared by the U.S. Department of Labor, Bureau of Labor Statistics, Office of Productivity and Technology, December 1979.

Policy Strategy to Stimulate Technological Progress

CED has concluded that new policies to stimulate innovation need to concentrate most on the problem of flagging
business investment. The chief causes of such flagging investment, in our analysis, are those that effectively reduce
the ratio of real rewards to risks incurred. Our study,
therefore, places most emphasis on remedial actions that deal
in a practical way with this shrinking ratio:

- Reducing Current Tax Disincentives from Out-moded
  Capital Recovery Allowances. Allowable depreciation
  on existing plant and equipment is based on historical
  costs, which in an inflationary period are much lower
  than replacement costs. This means that real cash
  flow is effectively held down while costs of new plant
  and equipment are going up. When the rate of return
  on investment is calculated on the basis of replacement cost, there appears a sharply diminishing incentive to invest in new plant and equipment. Since 1960,
  the real rate of return, based on replacement cost,
  has been substantially lower than the reported rate
  of return. In 1965, the real rate of return for U.S.
  nonfinancial corporations averaged about 10 percent
  while the reported rate was about 15 percent. By 1978,
  however, because of inability to control inflation and
  reliance on out-moded capital recovery rates, the real
  rate of return had declined to a little over 5 percent.

  There are a variety of measures available through the
  tax structure to make innovation investment more
  attractive to the private firm. These include increasing real rates of return on new projects (new facilities or improving existing facilities), expanding the
  fund flow from currently operating projects, thus making more capital available for new investment, and
  lightening the tax burden on the private sector.

  The Committee concluded that a more rapid capital recovery allowance deserves the highest-priority consideration by policy-makers. To accomplish faster
  capital recovery, CED recommended a phased introduction of accelerated capital depreciation rates. CED
  proposed that business be permitted to depreciate
  capital plant and equipment over fewer years than
  current law stipulates. Such a change would accelerate cash flow and reduce investor vulnerability to
  technological obsolescence and to inflation by allowing recovery of investment dollars of more equal purchasing power than is now possible. With more rapid

capital recovery, business will have greater incentive to invest in new plant and equipment which embody the fruits of research and development.

CED believes that such a policy change would stimulate commercialization of existing knowledge and diffusion of innovation throughout the economy. Remember that the change defers, rather than eliminates, investor tax liability. Over time, more rapid capital recovery provisions will generate new jobs, higher productivity and incomes, which will in turn more than offset the initial revenue loss for the U.S. Treasury.

- <u>Reducing the Risks and Uncertainties Created by the Current Regulatory Climate</u>. While recognizing that some government regulation is necessary, industry finds that in many cases compliance costs are badly out of line with any potential benefits. This reduces the remaining resources available for real technological advances. "Zero risk" goals, requirements for "best available technology," and the frequent changes in production standards sometimes required have all led to compliance cost escalation. The accompanying uncertainty about the acceptability of advanced technology applications has further thrown out of balance the risk/reward ratio for innovative investments.

The growth of regulations and the way they are implemented inhibits industrial innovation on a number of levels. First, regulation forces firms to reallocate their limited investment resources simply to comply with the mandated actions, thus diverting funds from research and development of new products and processes. Second, regulation increases the waiting time for return on investment, thereby discouraging investment in new, untried technologies. (The pharmaceutical industry has been especially hard-hit by this factor in recent years.) Third, many regulations increase uncertainty and reduce incentives to innovate because their standards are frequently open-ended and subject to change. This can make would-be investors extremely cautious about putting their capital into innovations.

CED believes there is an urgent need to determine if government action is required to solve a particular problem. If government action is needed, it should be invoked in a manner that least distorts the market system. This means, for example, substituting simple yet powerful economic incentives for detailed rules.

This would enable the accomplishment of the justifiable social purposes of regulation without unduly inhibiting innovation and the need for market flexibility.

CED also recommends that government reduce regulatory uncertainty by assessing each proposed regulation for its impact on innovation. Such an analysis would mean consideration of the potential effects on investment resources through restricting business's ability and incentive to apply capital to research, development, and innovation.

We applaud the efforts of the Administration and the Congress to achieve regulatory reform, and we particularly advocate the development of guidelines for determining whether new or existing regulations are actually needed, and the introduction of a broader and more systematic process for periodic review. In this way, the government may come closer to basing such decisions on the real costs and benefits of both social and economic regulations.

- Selective Tax Changes to Improve the Risk/Reward in R&D. There are a number of selective tax changes that would provide targeted help to earlier research phases of the innovation process. We believe the most worthwhile would be to allow complete flexibility in the choice of depreciation of capital expenditures for research and development uses. The current tax code permits a deduction for depreciation allowances for buildings and certain types of equipment used for R&D activities. CED would amend this to allow flexible depreciation for all such fixed assets. Accordingly, a taxpayer would have the option of depreciating R&D assets in the first year of their life or adopting any other method desired while retaining the benefits of the allowable investment tax credit. This type of flexibility exists under a number of the tax codes of our industrial competitors, including those of Canada and Great Britain.

The Committee also recommends consideration of some selective tax modifications to aid the innovative process. Faster depreciation for patents -- over a ten-year period or their demonstrated life, whichever is shorter -- is necessary, as the pace of modern technology makes many patents obsolete long before the end of their traditional 17-year life. Increasing the investment tax credit for R&D facilities from 10 to 20 percent would spur R&D expenditures. The government should consider allowing tax credits for corporate contributions to support nonproprietary research at universities. Pioneer plants, an especially difficult hurdle in the invention/innovation chain, could be encouraged through special tax measures.

Small business, traditionally the source of much innovation, has been especially hard-hit by excessive taxes and regulations. Government should help restore incentives for the smaller firm to innovate by: (1) increasing the deductibility of capital losses against ordinary income; (2) extending the carry-back provisions of the tax code for both net operating losses and investment tax credits; and (3) permitting Subchapter S corporations to raise capital from 100 investors rather than the present 15 maximum.

All of these selective tax changes, however, will not be as significant in their effects on innovation as faster capital recovery allowances. Indeed, these selective changes will be most effective if they are accompanied by the kind of general tax change recommended above.

- Reforming the Patent System. CED believes that the current patent system is unnecessarily complicated and unreliable. It also imposes substantial costs on business and the public, without adequately providing significant discoveries with the protection required to encourage innovation. Because patent laws make developed technology a controllable property, their effectiveness plays an important role in determining whether new-technology development will be undertaken at all.

An efficient patent system meets three vital criteria. The system should be accessible -- simple, inexpensive, and available to everyone. It should be reliable, allowing everyone to know the bounds of the protection and how well a given patent will stand up under litigation. Finally, a patent system must be selective; it should protect and encourage new knowledge without burdening the public with patents on minor and obvious variants of what was previously known.

In CED's view, certain aspects of the U.S. patent system fall short of the primary objective, namely, the provision of the effective incentives and protections for developing new inventions and encouraging

the utilization of the results of basic and applied research and development. To remedy this deficiency,
the CED study makes a number of recommendations, including:

- Allow voluntary arbitration to speed up the settlement
  of patent disputes. Major commercial patents often
  invite controversy. Industrial competitors frequently disagree about the scope or value of a
  patent, and protracted court proceedings result in
  high legal costs and years of delay. To reduce the
  costs and time involved in patent disputes, CED believes that arbitration should be available to
  those parties who wish to use it. Arbitration
  is common in many other areas of commerce, including labor settlements involving many millions of
  dollars, and we believe it would be similarly
  effective in resolving questions of patent infringe-
  ment and validity.

- Establish a single court of patent appeals to enhance uniformity in application of patent statutes.
  The establishment of such a court would go a long
  way toward eliminating the problem of inconsistency
  in precedents between existing U.S. Circuit Courts.

- Adopt a first-to-file patent system. This would
  strengthen the patent procedures and help to eliminate the extended delays that occur when two or
  more inventors claim the same invention. Under
  such a system (used by virtually all industrialized
  countries except the United States and Canada), the
  first inventor to file would be granted the patent.
  A personal right of use provision could be preserved
  for any inventor filing later who could show he invented first and took steps leading to commercialization.

- Liberalize granting of title to government contractors (business or academic) for inventions or
  patents developed under government contract. Over-
  zealous government restrictions in this area have
  markedly depressed commercialization of such innovations. Title to inventions made under these
  conditions should therefore be vested in the contractors, with the government retaining certain
  rights in the event of non-use of the patent.

- Improve current reexamination procedures by permitting defendants faced with adverse patent

claims to ask the U.S. Patent and Trademark Office for reexamination.

● Expanding Government Support of Basic Research.
Experience has shown us that government has generally not been very successful in picking "R&D winners." R&D to improve products and processes for market activities is therefore properly the responsibility of industry; industry is in the best position to make R&D decisions that must be coordinated with production and sales activities because it can judge better than government what is commercially viable.

Government does, however, have a role to play in developing innovations which are primarily for the government's own use. In programs such as defense or space research, or research into improved mail delivery systems, air traffic control, and sewage disposal, government is the primary consumer. As such, government is in the best position to judge R&D needs in these areas, for the same reason that the private sector has the best vantage point to assess R&D needs in areas of commercial application.

Another area in which the public sector must play an active part in financing is basic research. Basic research, much of which is carried on in the universities, can have enormous social payoffs. But, by its nature, fundamental research entails highly uncertain results and a high risk of failure. Although necessary as the first link in the technology chain, basic research is very expensive and so cannot be economically justified by the private sector alone. In addition, the nation as a whole gains from the education and experience acquired by the scientists and engineers engaged in basic research. Such talent is, in CED's view, an important national resource to be preserved and protected, and public support for the development of that talent is therefore justifiable.

The CED report on technology calls for the federal government to increase its relative funding of basic research, especially in universities, even at the cost of other activities. Other funding does in fact come from industry, foundations, and state budgets, but as a practical matter, there is no alternative to the federal government as the primary source of money. A high level of federal support should be explicitly stated as a major goal of national policy. The prospect of stable funding has become increasingly vital as research costs at universities have risen steeply in the last decade. In the long run, a successful basic research program will furnish the nation with a strong base of knowledge upon which future economic growth and industrial innovation can be founded.

To bolster basic research capabilities, the CED report urges the federal government to give the highest priority to increasing technical and scientific instrumentation in universities, while allocating some funds to renovating obsolete research facilities. The public sector should, in addition, adopt a system of full cost reimbursement for university research done under government contract, eliminating the need for academic institutions to divert non-research resources to maintain such programs.

CED also urges that the federal government place greater emphasis on supporting the work of outstanding scholars and less emphasis on equal distribution of funds among a large group of institutions. Finally, CED believes the government ought to foster joint research programs between the business and academic communities, perhaps by allowing business to claim tax credits for financing of such programs.

Followup to the CED Study

We at CED have been pleased by the initial reaction to Stimulating Technological Progress. Extensive coverage of the report's recommendations was carried over the major wire services, in major national newspapers, and in the trade and education press. Much of the coverage focused on CED's contention that the United States' ability to deal with unemployment, inflation, and competition from foreign industrial powers is being eroded by public policies that discourage innovation. An editorial in Electronics Magazine called attention to the study, saying that U.S. policy toward technology and innovation will "establish the international economic position of the nation in the 1980's and 1990's: Its standard of living, its ability to deal with social problems, and its defense posture." An Indianapolis Star newspaper editorial praised the report and said that the measures recommended by CED could "dramatically improve productivity, the level of capital investment and the economy as a whole -- including the availability of jobs for American workers."

The report is being very well received on Capitol Hill.

CED representatives have already testified before a number of
Congressional committees about the importance of stimulating
technological progress. In February 6, 1980, testimony,
before the Joint Economic Committee of Congress, Franklin A.
Lindsay, Chairman of CED's Research and Policy Committee
and Chairman of the Itek Corporation, spoke about the need
for faster recovery of capital invested in R&D-intensive
plant and equipment:

> Tax incentives to encourage more capital invest-
> ment are urgently needed as a way to improve
> productivity and fight inflation over the long
> run. It makes no sense to say that we cannot
> adopt such long-term structural measures unless
> there is a recession. Indeed, by delaying the
> introduction of improved depreciation allowances,
> we will fall further behind in our efforts to
> produce the new capital -- and prevent the deteri-
> oration of existing capital -- needed to avoid
> major inflationary bottlenecks in the future....
>
> A strategy for recapitalizing the American economy
> should also include a range of measures to induce
> higher levels of private investment in longer-term,
> and often high-risk, R&D leading to innovation and
> its diffusion throughout the economy. The indus-
> tries most directly benefited by a stepped-up and
> sustained R&D include those on which the nation
> will increasingly depend for efficiency, inter-
> national competitiveness, rising real income, and
> national security.

The links between capital investment, innovation, and
productivity improvements were underscored by Thomas A.
Vanderslice, Chairman of CED's Subcommittee on Technology
Policy, in testimony before the House Budget Committee on
February 21. Mr. Vanderslice, who is also President of
General Telephone & Electronics Corporation, pointed out
that Japan and West Germany, two of our major industrial competitors, are renewing the industrial base of their economies
at twice or more the rate of the United States. The United
States, by contrast, is beginning to experience what he termed
"industrial hardening of the arteries," with a leveling off
of productivity growth.

To reverse this trend, Mr. Vanderslice urged the Budget
Committee to give top priority to tax measures to increase
capital investment. But he added, "if we wish to sustain
the productivity growth stimulated by increased capital

investment we must encourage industry to maintain and increase its relatively high level of investment in research
and development throughout the 1980's. Mr. Vanderslice concluded by recommending "bold action" to spur capital investment, and said that the need for incentives for investment is
so great that Congress must act even by reducing expenditures
if necessary.

On February 8, before the House Subcommittee on Science,
Research and Technology, Franklin Lindsay underscored the
need for a more consistent patent system to facilitate the
utilization of government-funded research and development.
He pointed out that the present "confusion and uncertainty
leads to a disincentive to firms which may wish to invest the
funds necessary to commercialize a federally developed invention." He concluded, "In the process of trying to keep contractors from profiting from ideas born in the course of
government contracts, we are probably preventing the development and distribution of innovations that could benefit all
of society."

Mr. Lindsay's patent testimony was reinforced by that of
another CED representative, who testified against a proposal
that contractors be required to pay high fees to the government for use of a patent developed under a government contract. He pointed out that the proposed payments will act as
a disincentive to commercialization which the federal government wants to promote. Calling it "nothing more than a charge
or tax," he added that the payment proposal will "raise the
cost of the new products and make it harder for them to compete, particularly against foreign competitors in the world
market."

To carry the message of CED's study on technology to
other forums, CED will be holding joint discussions and meetings with organizations that represent the research and development communities, both in the academic and the industrial
worlds. CED is confident that a thorough airing of the issues
involved in stimulating innovations will foster greater and
more productive joint efforts between the research communities,
and help to inform the policymakers in Washington.

Conclusion

Like my colleagues, I believe that the problems of sluggish productivity and less than desired innovation in the
United States are primarily structural. Therefore, the solution must lie in inducing the needed structural change in the
productive and innovative base of the economy.

CED's proposed strategy for spurring technological innovation can be summed up as follows:

- Begin by improving the incentive for creating new production facilities. If we can raise the level of investment in plant and equipment, we will increase immediately the rate of diffusion of new technology into the economy and improve the rate of productivity growth.

- For measures to achieve this, our first priority would be reduction of existing tax disincentives to productive investment, specifically by allowing faster depreciation of new plant and equipment.

- The next priority would be the reduction of non-cost-effective regulatory constraints and uncertainties.

- Next, appropriate patent, tax, and regulatory changes should be made to provide support to foster private research and development.

- In addition, adequate support of basic research at universities should be a high-priority item in the federal budget.

In our judgment, these reforms should induce the physical investment necessary for a lasting improvement in productivity, thereby contributing to the long-run dampening of inflation. In conjunction with the improved economic performance and increased demand for advanced technology that would result from a higher level of productive investment, the above policy actions should help create a climate in which outlays in all phases of technological innovation would be increased as a natural result of the entrepreneurial process.

This strikes the members of CED's study group as the surest path to technological progress.

Appendix A

Subcommittee on Technology Policy in the United States

Thomas A. Vanderslice, Chairman
President
General Telephone & Electronics

David Beretta
Chairman
Uniroyal, Inc.

John C. Bierwirth
Chairman
Grumman Corporation

Derek Bok
President
Harvard University

Alfred Brittain III
Chairman
Bankers Trust Company

Fletcher L. Byrom
Chairman
Koppers Company, Inc.

Robert D. Campbell
Chairman of the Board
Newsweek, Inc.

William S. Cashel, Jr.
Vice Chairman
American Telephone and
    Telegraph Company

Richard M. Cyert
President
Carnegie-Mellon University

John Diebold
Chairman
The Diebold Group, Inc.

Harry J. Gray
Chairman and President
United Technologies Corporation

Frederick G. Jaicks
Chairman
Inland Steel Company

Edward R. Kane
President
E.I. duPont de Nemours & Co.

Charles N. Kimball
Retired President
Midwest Research Institute

Jean Mayer
President
Tufts University

C. Peter McColough
Chairman
Xerox Corporation

Thomas O. Paine
President
Northrop Corporation

D. C. Searle
Chairman, Executive
    Committee
G. D. Searle & Company

Robert B. Semple
Chairman
BASF Wyandotte Corporation

Mark Shepherd, Jr.
Chairman
Texas Instruments, Inc.

David S. Tappan, Jr.
Vice Chairman of the Board
Fluor Corporation

Subcommittee, continued	Nontrustee Members
Howard S. Turner Chairman, Executive Committee Turner Construction Company	Norton Belknap Senior Vice President Exxon International Corporation

George L. Wilcox
Director-Officer
Westinghouse Electric
    Corporation
	Jerrier A. Haddad
	Vice President - Technical
	    Personnel Department
	IBM Corporation

J. Kelley Williams President First Mississippi Corporation	Gerald D. Laubach President Pfizer, Inc.

Richard D. Wood
Chairman of the Board
    and President
Eli Lilly and Company
	George M. Low
	President
	Rensselaer Polytechnic
	    Institute

Advisors to the Subcommittee

Richard L. Garwin
IBM Fellow and Science
    Advisor to the Director
    of Research
John Fitzgerald Kennedy
    School of Government
Harvard University

Tait S. Goldschmid
Senior Economic Analyst
Exxon Corporation

Walter Hahn
Senior Specialist, Science
    and Technology and
    Futures Research
Congressional Research
    Services
Library of Congress

J. Herbert Hollomon
Director, Center for Policy
    Alternatives
Massachusetts Institute of
    Technology

Ronald M. Konkel
Office of Management and
    Budget (on leave)

Wesley A. Kuhrt
United Technologies Corporation

Max Magner
Staff Consultant/Technical
    Government Liaison
E.I. duPont de Nemours and
    Company

Harry Manbeck
General Patent Counsel
General Electric Company

Boyd McKelvain
Staff Associate - Technology
    Policy Development
General Electric Company

Rudolph G. Penner
Resident Scholar
American Enterprise Institute
    for Public Policy Research

Advisors, continued

Rolf Piekarz
Senior Staff Associate
Division of Policy Research
    and Analysis
National Science Foundation

Roger Seymour
Program Director, Commercial
    Relations
IBM Corporation

Project Director

Edwin S. Mills
Chairman, Department of
    Economics
Princeton University

CED Staff Counselor

Kenneth McLennan
Vice President and
    Director of Industrial
    Studies

CED Staff Advisor

Frank W. Schiff
Vice President and Chief
    Economist

Project Editor

Claudia P. Feurey
Director of Information
Committee for Economic
    Development

Project Staff

Lorraine Mackey
Administrative Assistant

References

(1) <u>Stimulating Technological Progress</u> (New York: Committee
     for Economic Development, January 1980); available from
     the Committee for Economic Development, 477 Madison
     Avenue, New York, N.Y.   10022. Paperbound, $5.00 plus
     $.50 postage, prepaid.

(2) See CED's policy statements, <u>Fighting Inflation and
     Promoting Growth</u> (August 1976); <u>Jobs for the Hard-to-
     Employ: New Directions for a Public-Private Partnership</u>
     (January 1978); <u>Helping Insure Our Energy Future: A
     Program for Developing Synthetic Fuel Plants Now</u> (July
     1979); <u>Redefining Government's Role in the Market System</u>
     (July 1979); and CED's supplementary paper, <u>Thinking
     Through the Energy Problem</u> (March 1979).

(3) A variety of scholarly studies have been inspired by
     Edward Denison's comprehensive 1962 study, <u>The Sources of
     Economic Growth in the United States and the Alternatives
     Before Us</u> (New York: Committee for Economic Development,
     January 1962).

(4) Roger Brinner and Miriam Alexander, <u>The Role of High-
     Technology Industries in Economic Growth</u> (Cambridge,
     Mass.: Data Resources, March 1977).

(5) National Science Board, <u>Science Indicators</u>, 1976 (Wash-
     ington, D.C.: National Science Foundation, 1977),
     Table 1-23. Technological intensity is measured by the
     relative amount of research and development performed by
     the pertinent businesses.

_____ *Walter A. Hahn, Mary Ellen Mogee*

5. Research and Innovation

Introduction and Background

This paper describes the Research and Innovation Area Study (RIAS), which was recently completed for the Joint Economic Committee (JEC) of the U.S. Congress. It describes the origin and objectives of the study and summarizes the findings and conclusions. Supporting detail may be found in the full report which will be available from the JEC.*

The JEC is a nonlegislative committee of the Congress, established in 1946 as the congressional counterpart to the Council of Economic Advisers. It was intended and has functioned as one of the major governmental institutions involved in formulating Federal economic policy. The JEC does not introduce legislation. Instead, its traditional role has been one of information and education. It has been influential through its study activities and through the overlap of its membership with other legislative committees. Senator Lloyd Bentsen (D. Tex.) is currently chairman of the JEC. The JEC is in a unique position to address central economic issues that require a long-term perspective and broad outlook. For this reason, RIAS represented an excellent opportunity to integrate scientific and technological considerations into the examination of economic policy.

The RIAS was done as part of a major reexamination of the structure of the U.S. economy and the role of Federal policies, called the Special Study of Economic Change (SSEC). Since 1946, when the JEC was established, the U.S. economy has changed

_____
* U.S. Congress Joint Economic Committee, Special Study on Economic Change, Vol. 3, Research and Innovation. Studies prepared for the use of the special study on economic change of the Joint Economic Committee, Joint Committee Print, Washington, U.S. Government Printing Office, 1981.

dramatically. Not only have economic conditions changed, but also, it seems, the basic forces which structure the economy. Traditional neoclassical economic theory has proven incapable of explaining or predicting recent economic changes. Hence it no longer appears to be adequate as a basis for public policy. The SSEC was an effort to prepare for the future by determining the major sources of economic change in the remainder of the 20th century.

The SSEC was approved by the Congress in July 1977 under the leadership of then JEC chairman Rep. Richard Bolling (D.-Mo.). It was conceived as a study effort to produce information and analysis for consideration by the members of the JEC that might eventually result in the formulation of legislation and its introduction by other committees.

The SSEC consisted of ten area studies, of which research and innovation was one. Together, these ten areas constituted a broad look at major sources of change in the U.S. economy. The ten areas were: (1) human resources and demographics; (2) materials and energy; (3) stagflation; (4) Federal sector finances; (5) State and local finances; (6) pension systems; (7) government impact; (8) international environment; (9) productivity; and (10) research and innovation. At the time of this writing, it is planned that each of the ten area studies will be published by the JEC as staff reports. These will be published during the summer and fall of 1980. In addition, the Joint Economic Committee will publish its own summary report on the entire project, with its findings and recommendations late in 1980.

Industrial Innovation as a Public Policy Issue

The RIAS attempted to document what is known about the role of technological innovation in the economy, past trends and the present state of the innovation "system," and the outlook and some congressional options for the future. Technological innovation was defined for these purposes as the process by which new and improved products, processes, and services are generated and come into use throughout society. Technological innovation includes the activities ranging from idea generation through research and development (R&D), commercialization, diffusion, adoption, and use. The name Research and Innovation Area Study was chosen to emphasize the idea that technological innovation includes R&D but goes beyond it.

Technological innovation in industry, or industrial innovation, has become a public policy issue of increasing concern in recent years. Articles have appeared in popular magazines with such alarming titles as "Vanishing Innovation" and "The Breakdown of U.S. Innovation." On the eve of the nation's two hundred and second birthday, the lead sentence in a Business Week article on "Vanishing Innovation" read:

> A grim mood prevails today among industrial research managers. America's vaunted technological superiority of the 1950s and 1960s is vanishing, they fear, the victim of wrongheaded Federal policy, neglect, uncertain business conditions, and shortsighted corporate management.

General Electric's chairman Reginald H. Jones in News-week summed up the dominant view, "There's a new mood in America that could bring the grand enterprise of science and technology to an ignominious halt." The article went on to add that "lately, technocrats have begun to fear that America's lead in innovation -- and the vaunted U.S. technological superiority that it spawned -- may be withering."

On the other hand, Vice President John J. Wise of Mobil Research and Development Corporation saw it differently. "There is an enormous amount of technological development going on in the country. I think the state of innovation is much better than most people are aware." Likewise, House Science and Technology Committee chairman Don Fuqua stated, "Fundamentally, I do not believe the United States has lost its innovative spirit that has enabled us to maintain our preeminent economic position in the world." He went on to say, however,

> I am convinced that deliberative, coordinated actions must be taken by Congress and the administration to remove disincentives to innovators and to help provide a framework that will help us manage our inventions better. Furthermore, actions to improve our innovative capacity must be part of an overall strategy to improve our sagging productivity performance.

On October 31, 1979, President Carter announced a series of initiatives to stimulate U.S. industrial innovation. These initiatives followed a lengthy domestic policy review on industrial innovation in the executive branch which involved inputs and deliberation by high-level officials in both Government and industry, as well as labor and public interest representatives. President Carter opened his message to the Congress with these words:

> Industrial innovation -- the development and commercialization of new products and processes -- is an essential

element of a strong and growing American economy. It
helps ensure economic vitality, improved productivity,
international competitiveness, job creation, and an
improved quality of life for every American. Further,
industrial innovation is necessary if we are to solve
some of the Nation's most pressing problems -- reducing
inflation, providing new energy supplies and better
conserving existing supplies, ensuring adequate food
for the world's population, protecting the environment
and our natural resources, and improving health care.

Our Nation's history is filled with a rich tradition
of industrial innovation. America has been the world
leader in developing new products, new processes, and
new technologies, and in ensuring their wide dissemination and use. We are still the world's leader. But
our products are meeting growing competition from abroad.
Many of the world's leading industrial countries are now
attempting to develop a competitive advantage through
the use of industrial innovation. This is a challenge
we cannot afford to ignore any longer.

The rest of this section attempts to provide a synoptic
view of the industrial innovation policy area as it has developed in the last few years. "Innovation" has emerged in
the last two years as a hot national topic. Technologists
and businessmen see it as being stifled, or at least retarded,
by regulatory and "micro-management" barriers. Our national
innovation "elan" is said to be gone. Also "gone," in the
view of some, are sufficient rewards for entrepreneurial risk-
taking and a stable environment for initiating long-range
endeavors. Those concerned with improving economic affairs
are beginning to recognize innovation as an essential element
in the satisfactory performance of the domestic and global
economies. They also are becoming aware that the health of
the economy is a major factor in the health of innovation.

At the same time, those who study the innovation process
and its role in socioeconomic activities point to a lack of
awareness and use of much of what is known. There is a continuing need for more appropriate theory and for reliable,
meaningful data. Consumer, environmental, and labor groups
have varying concerns for the negative impacts of present
levels and direction of innovation. They worry that innovation enhancement will be used as an argument to repeal hard-
won protective measures so as to "enhance profits."

These groups are making competing claims on Government.
The pressures on politicians to resolve these differences are
increasing much faster than clear and accurate knowledge upon

which to base action and to assess consequences. Additionally, those who must resolve these issues find themselves bound by the precedents and constraints of public and private institutions designed for earlier times and situations.

A complicating factor is the lack of consensus as to what the innovation "problem" is. This may be because there are actually many interrelated problems -- and opportunities. Another reason may be the complexity and interactive nature of all of the elements of the problem(s). This applies both within what this study called the "innovation system" and with the two-way relationships between the innovation system and the larger set of national and global social, economic, and political affairs.

In addition, there is a debate as to whether or not there is an objective innovation problem(s). The evidence bearing on this question is mixed. Although trends in a number of indicators show declines in the United States relative to earlier years and to certain other industrialized nations, these data are not universally accepted. The "innovation problem" may exist mainly in the subjective perception of certain people. For public and private policymaking purposes it does not seem to make much difference what the answer is. If industry, Government, consumer, and other group representatives believe things are not going well they will behave accordingly. If that behavior hinders the achievement of social and economic goals, Government officials will be pressured and have a responsibility to deliberate and possibly to act.

It is also clear that the relationship between Government and industry in the United States has developed some negative and confusing overtones. Some industries want less "interference" in regulatory affairs, others want selective protection from foreign competitors, while still others seek Government bail-outs. International traders simultaneously want more freedom ("Let's not export our antitrust and bribery laws to other cultures") and more help ("Why can't we be more like Japan Incorporated and have our Government behind us in international competition?") Serious misunderstandings, ignorance of each other, and perpetuation of old or false myths characterize relationships between industry and Government. Mutual trust and respect appear generally to be very low. The actions of many of the parties in coping with each other come as surprises, which in turn generate reactionary responses, often with negative consequences for all. Defensiveness, the search for "blame," and an ever-shortened outlook abound.

It was within this policy issue context that the RIAS was conducted. The next section describes the RIAS final report.

The RIAS Final Report

Studies of innovation are usually based on an explicit analytical model. None of the dozens of models used in prior innovation studies, however, seemed adequate for this broad and far-sighted study of a complex subject. The only model consciously used in RIAS was the systems model. In conventional systems analysis fashion, the elements of the innovation system were dealt with separately and in the context of their relationships to each other. Simultaneously, innovation was conceived as a subsystem of the larger economic and social whole, both influencing and being influenced by the larger system. The one overriding theme, perhaps even conclusion of RIAS, is the necessity of viewing innovation in the systems context described above.

It was concluded early that major new research on technological innovation was not needed for SSEC purposes. It was judged to be more useful to review and synthesize the numerous past studies in this field. In a few areas where current thinking has gone beyond the available literature new papers were commissioned.

The RIAS final report was designed for use in several contexts: in the analyses and option reviews of the Special Study on Economic Change; by the Members of Congress in the legislative, investigative, oversight, and foresight duties; and by interested members of the public. In addition to the prepared papers, selected reprints and abstracts have been included in appendices. The report provides many of the materials needed for further policy analyses and decision-making on the role of innovation.

RIAS emphasized what has been learned in the last two decades, reviewed current analyses and actions, and identified the major societal and economic impacts of and on innovation that are likely to occur in the next three decades. Three time frames are explicitly treated in the RIAS report. The first time frame is the past -- going back two decades. This section describes the state of knowledge about technological innovation and its role in the economy, as well as past policy studies and their recommendations. The second time frame is the present. It includes recommendations from current policy studies, current legislation, the results of a RIAS workshop on current innovation research and analysis, as well as the specially commissioned papers on emerging issues and concepts relevant to technological innovation. The third time frame extends three decades into the future with an emphasis on the middle 15 years. This section identifies the future major trends in technological innovation and the social and economic factors that will likely interact with technological innovation in the next 30 years. Brief synopses of the papers in the final report follow.

A. What is Known?

The first section of the RIAS report addresses the question of "what is known" about industrial innovation and the economy. A key paper describing the current state of knowledge is: The Process of Technological Innovation in Industry: A State-of-Knowledge Review for Congress. One thesis of this paper is that, although knowledge remains limited, recent (e.g., in the past decade) research has resulted in more information about the process of industrial innovation than most policymakers realize. The objective of the paper is to summarize and translate the research findings into a form useful to congressional staff and Members.

The report's findings have implications for congressional decisionmaking for innovation policy. Listed here in brief form, they are discussed in more detail in the report:

- Innovation is a complex process and our understanding of it is limited;

- The essence of innovation is uncertainty about the outcome;

- The importance of market factors to industrial innovation is difficult to overemphasize;

- Innovation is a costly and time-consuming process;

- The economic and social impacts of innovation are made through their diffusion;

- Basic scientific research seems to underlie technological change in complex and indirect, but important ways;

- The innovation process differs from industry to industry, sector to sector, and even firm to firm;

- Financial and manpower resources are necessary, but not sufficient, for innovation; and

- Both large and small firms play important roles in innovation, and those roles differ from industry to industry.

In parallel with academic research on innovation have
been a plethora of policy and issue studies by a variety of
boards, conferences, contractual and in-house analytical teams,
and a few individuals. In the chapter entitled Two Decades of
Research on Innovation: Selected Studies of Current Relevance
are collected many proposals for policy and action that may be
still relevant but are largely unevaluated and untried. One
example is "The Charpie Report" of 1967 which is still widely
cited as a nonimplemented "classic" in this field. The 205
recommendations contained in these 42 reports are extracted
and grouped for ease of assimilation. A consolidated display
of all recommendations attempts to show the action parties to
which the recommendations appear to have been directed along
with those groups who would be impacted.

Taken together, these selections provide the main sources
for much of the current understanding of the innovation system
and illustrate the many interconnected issues which must be
faced by policymakers seeking to stimulate technological innovation. Likewise, the identification of action parties connected with the recommendations indicates how many segments
of the Federal Government intersect with the innovation system
and suggests the need for integrated and comprehensive policymaking. No in-depth analysis of the individual recommendations, however, has been attempted.

1. Domestic Policy Review Advisory Committee Report and
    President's Message on Industrial Innovation

This chapter provides a brief summary and analysis of the
Final Report of the Advisory Committee on Industrial Innovation
and the President's Message on Industrial Innovation. It permits a comparision of the "outputs" of the Domestic Policy
Review -- that is, the President's recommendations -- to the
public "inputs" to the Domestic Policy Review -- that is, the
policy recommendations of the Advisory Committee. Such a
comparison is of interest because of the numerous complaints
by industry that their recommendations were not heeded and
criticism by industry and Members of Congress that tax incentives were not included in the President's Message.

In parallel with the executive branch, the Committee for
Economic Development (CED) conducted an in-depth policy
analysis entitled Revitalizing Technological Progress in the
United States. The CED summary chapter is reprinted in the
RIAS report with CED's permission. The CED proposed a strategy for technology in the U.S. economy which consisted of two
principal elements. The CED recommended that, first, the
level of investment in plant and equipment should be raised
in order to increase the rate of diffusion of new technology

into industrial processes. Second, CED called for reducing
"unessential" regulatory constraints and uncertainties on
productive investments. Other policy changes were recommended
in the areas of patent policy and direct Federal support for
R&D.

> 2. The Relationship of Federal Support of Basic Research
>   in Universities to Industrial Innovation and Pro-
>   ductivity

Another chapter in the "known" category presents an over-
view of what is known about the relationship of Federal support of basic research in universities to industrial innovation and productivity. It reviews three kinds of evidence
which bear upon this issue: the conceptual relationship between science and technology, the nature of university-industry
relations, and economic studies of the contribution of research and development to economic growth and productivity.

The report reveals that there is widespread agreement
among university, Government, and industrial officials that
Federal support of basic research in universities is an
effective method of enhancing the science base for industrial
innovation. However, economic studies have been unable to
isolate the precise quantitative contribution of basic research (as opposed to applied research and development) to
economic growth and productivity. The existence of institutional barriers between universities and industry may be
obstructing the transfer of basic research results to industry
and preventing them from being embodied in new technology
which is necessary in order to contribute to improved economic
productivity.

The report concludes that Federal funding of basic research in universities may be viewed as an investment that
will have payoff primarily in terms of improved efficiency of
the R&D process and major technological changes that may permit continued improvements in economic productivity in the
long-term future.

B. Status

Realizing that the cutting edge of research and practice
may be ahead of much of the research and policy analysis
literature, an attempt was made in the second section of the
RIAS report to update, and to some degree authenticate, the
picture presented in the first section. For example, Research,
Innovation, and Economic Change is a summary and analysis of
the presentations and discussions of a December 1978 "synthesis" workshop. Among the themes stressed during the workshop

discussion were (1) the apparent decline in the validity of U.S. technological infrastructure and the loss of an "innovative elan," and (2) the lack of definitive knowledge about the relationships between investments in research and development and desirable economic change. Although the workshop discussions produced no consensus on specific policy options, the general sense of participants was that: (1) there was a need for both remedial and anticipatory policy actions in the research, development, and innovation areas; and (2) there is enough knowledge, given the risk of no action, on which to base policy choice. Policy options identified in the workshop report as meriting particular attention include:

1. developing attitudes and mechanisms supportive of positive Government-business relationships in the areas of civilian research and industrial innovation;

2. examining the organizational structure of the executive branch, with respect to its ability to carry out the Federal role in those areas, including the support of basic and applied research for industrial application;

3. identifying existing Federal policies and practices which act as barriers or deterrents to innovation, and where it is possible without compromising the primary objectives of those activities, modify them to remove or reduce their negative innovation impacts;

4. lessening congressional pressure (or at least correcting the perception of such pressure) for short-term evidence of the success of Federal actions in support of industrial innovation, including research support; and

5. developing incentives for labor and labor unions to accept, if not actively support, technological changes in the manufacturing and service sectors.

   The workshop focused intently on the operation of the innovation system -- it was an inward look. A parallel view from the outside, from "the economy" so to speak, is presented in Technical Advance and Economic Growth: Present Problems and Policy Issues. This paper analyzes the cause of some of the aspects of the present economic malaise such as inflation, unemployment, and declining productivity growth. It concludes that the significant deceleration since 1973 in R&D expenditures has been primarily due to the deceleration in growth of economic output. Slow and conservative technical advance can make it more difficult to get out of the current economic rut; while faster and more innovative technical advance may make it easier to get out. The author does not advocate Government

stimulus of basic technology as the most important instrument
in resolving today's macroeconomic problems, but argues that
such policies can be important parts of an effective policy
package.

To assure an up-to-date and comprehensive review of the
status of legislative action related to innovation, the RIAS
report also contains the latest Congressional Research Service
"Issue Brief" entitled Industrial Innovation. It records
major relevant bills and other legislative actions, along with
statements on the issues and actions involved and includes
appropriate supporting and background materials.

Thus far, the reader has been offered multiple views of
the present status of innovation in the United States. They
range from perceptions from within the innovation system, from
the "economy," a focused view of congressional activities, to
a comprehensive survey (though notably incomplete) of the
plethora of ongoing innovation studies, reports, meetings,
seminars, and colloquies. For a different synthesis of much
that has been said above, placed in balanced perspectives both
in terms of time and the international milieu, one additional
"status report" was written from the viewpoint of an entrepreneur. In The Revival of Enterprise, the author defined
enterprise as "the willingness to venture on bold, hard, and
important undertakings with energy and initiative. He concluded:

> Yes, enterprise did once flourish in our country and has
> now diminished in intensity. The reasons for its current
> lack of vigor are hard to pin down exactly, but include
> satiation, diversion of resources from "productive" to
> "unproductive" pursuits, and above all, increased uncer-
> tainty occasioned by inflation and regulation. Uncer-
> tainty results in short-term perspective. There are
> many proposals for stimulating innovation and productivity,
> but their effectiveness is not certain, by any means.
> Even if these proposals work as intended and they stimu-
> late innovation and productivity, they may not rekindle
> enterprise. After all, innovation may be channeled to
> trivial pursuits and our definition of enterprise requires
> hard, bold, and important action. Finally, the opportu-
> nities and needs for enterprise abound and in some in-
> stances, at least, there seem to be few alternatives.

1. Innovation Policy Recommendations from Selected
    Current Studies

The final chapter in the "status" section is largely a
follow-up to the earlier chapter, Two Decades of Research on

Innovation. In this case, however, the studies were all completed in 1978, 1979, or 1980. As in the earlier chapter,
policy recommendations were extracted, categorized, and displayed in relation to possible action parties. In addition,
there is an annotated bibliography and selected summary ex-
cerpts from the studies. This chapter serves at least two
purposes. By itself, it permits rapid scanning and quick comparison of the policy recommendations that have been made in
the latest round of the debate on industrial innovation. Used
in conjunction with the earlier chapter on Two Decades of
Research on Innovation, it permits a comparison of current
policy recommendations with those of previous years. In
general, it appears that many of the earlier recommendations
are perceived by students of innovation and policymakers
alike to be of continuing relevance.

C. Explorations

     Several topics are current within innovation circles
that are inadequately dealt with in the existing literature.
Five papers were commissioned to give them systematic and in-
depth attention. One is entitled Science Indicators: Improvements Needed in Design, Construction, and Interpretation.
This is an evaluation by the General Accounting Office of
measures currently used to assess the status of U.S. science
and technology with particular attention to the needs of
policymakers. The biennial National Science Board's Science
Indicators reports (especially the most recent one at the time
of writing, Science Indicators -- 1976) were reviewed because
they contain measures which attempt to portray the significant
changes in the state of science and technology. Such measures
are potentially a valuable resource in Federal decisionmaking.
GAO examined the indicators in Science Indicators -- 1976 to
determine their validity, their limitations, and possible
improvements in their selection, design, and interpretation.
(Since this analysis the National Science Board has released
the newest edition entitled Science Indicators -- 1978.)

     The GAO made several recommendations for improving
Science Indicators: The National Science Board and the
National Science Foundation staff should continue to experiment in the Science Indicators series by developing and testing new indicators. They should emphasize a more conceptual
approach which first identifies what will be measured and
then generates the appropriate data. Attempts should be made
to develop indicators of the process and substance of research
and to better differentiate between science and technology.
More interpretation of the indicators should be included in
future reports.

The National Science Board and the Foundation agreed with some of the GAO recommendations, including the need for inter- pretation of the indicators and stated that several changes consistent with the recommendations were made in Science Indicators -- 1978. The Foundation disagreed that there should be more emphasis on a conceptual approach. Addition- ally, NSF believed that the present Science Indicators reports appropriately separate science from technology.

Another paper was commissioned on The Role of Imbedded Technology in the Industrial Innovation Process. Roughly defined, the paper focuses on that great bulk of minute, incremental technological changes and advances that constantly occur in all manufacturing, maintenance, and operational activities throughout the technological infrastructure of an industrial society. This has been termed imbedded technology, or "IT," and concerns a multitude of tiny advances not direct- ly resulting from planned R&D efforts. A key part of the effort was to define imbedded technology and to describe its nature and extent with particular emphasis on its critical, but often unrecognized role in innovation.

One of the problems in analyzing and assessing policy for innovation is classifying the technology and/or its appli- cations in consistent and measurable ways. One approach to the problem is offered in an exploratory paper, A Quantitative Technology Index to Aid in Forming National Technology Policy. This paper attempts to go beyond the widely used but vague terms, "high" and "low" technology, to provide an extended and more replicable method of categorizing technology for policy- making purposes. The index proposed is composed of multiple subjective scales in three descriptive areas: the technologi- cal product per se, the process of its manufacture, and the nature and extent of the distribution system. As with the other commissioned papers, this is a "think piece" outlining concepts and approaches but stopping short of development of a working tool. If a continuing need is perceived and if this approach appears to have merit, a subsequent action could be to initiate development and testing of the approach.

Most of the attention of the literature and current dis- cussion is directed toward our American style of large-scale, sophisticated innovation. This is perhaps as it should be, for therein has lain the major effort and resource commitment, the largest benefits, and the most critical policy and pro- cedural questions of the past. The chapter entitled The Role of Small-Scale Technology in Innovation deals with an issue just emerging on the American scene -- innovation for small- scale, decentralized, low energy, low pollution, and possibly more labor-intensive technologies and processes. As yet we are unaware of the full nature and extent of this area and thus of its policy implications. This chapter supplies a view of this parallel and future oriented counter-trend.

Another emerging topic is <u>Innovation in Public Technology.</u> A more accurate title for this chapter might be "the lack of innovation in the (so-called) public technology area." Public demands and Federal regulations are placing increasing responsibilities on state and local governments. Concurrently, budget limitations are constraining the amount of resources that these jurisdictions can spend meeting the needs of their citizens. One solution to this dilemma is to increase the productivity and effectiveness of public goods and services delivery through the application and utilization of technology. This process -- labeled "public technology" in the state and local sector -- affords a mechanism to foster new innovation to supply solutions to state and local problems. The state and local marketplace for innovation, however, is characterized by policies, practices, and organizations which, when combined with an apparent lack of technical expertise, tend to lead to a "no-risk" environment. Further, because of the absence of an aggregated market, industry has tended to avoid participation in the public technology venture. To fill the gap between what states and localities need and what technological solutions are available, the Federal Government has created various technology transfer and technical capability building programs. Yet some argue that industry could be attracted to innovation in the state and local arena if public technology markets could be identified and aggregated.

D. <u>Outlook</u>

Relating to the past and emerging from the present are a mix of future issues and opportunities relating to science and technology. A central chapter in the RIAS report is therefore <u>A Science and Technology Outlook</u>. Covering a period starting about five years in the future, this outlook goes out three decades, with the emphasis on the 15-year middle zone. The <u>Outlook</u> identifies those factors internal and external to science and technology (S and T) that need to be understood by policymakers to integrate S and T policy effectively into the overall techno-economic policies of the Government. In one sense, this longer range outlook picks up where the executive branch <u>Science and Technology Five Year Outlook</u>, mandated by the National Science and Technology Policy Organization and Priorities Act of 1976 (P.L. 94-282), ends. This initial study is an attempt to set forth a preliminary structure for a comprehensive science and technology outlook, par-

ticularly as science and technology relate to economic change.
As part of this outlook, a number of recent studies of the
future of science and technology, or of some selected aspects
of the future of science and technology, were reviewed for
methodology (or approach) and content.

The approach is fivefold. It is to examine the systemic
factors involved in the Nation's scientific-techno-economic-
sociopolitical system; investigate the total systemic "environment" in which the national science and technology system
operates; identify emerging technological developments which
seem likely to be important in the near-term future; review
basic assumptions; and analyze the preliminary findings and
suggest policy alternatives.

Considering basic assumptions first, three scenarios
incorporating alternative sets of basic assumptions are developed. The "extrapolative" scenario now seems to be the
most likely and most preferable of the three presented in
this analysis, although it includes significant existing and
emerging problems. Hence, national science and technology
policy alternatives are likely to be developed around this
basic scenario, or a similar one, either implicitly or explicitly. The thrust of the extrapolative scenario is that
the principal parameters of the global (or, at least, American)
socio-politico-techno-economic system will remain fairly constant over the immediate and near-term future, that is, for
at least the next 30 or so years. Another way of saying this
is that the future will be mainly characterized by extrapo-
lations of existing trends. Two alternative scenarios considered for comparison are the "changing values" and "dis-
continuity" scenarios.

Assuming the general validity of the extrapolative
scenario, the national science and technology system may be
called upon to contribute, in a global context, to the solution
of most, if not all, of the 14 major world problem areas discussed in the report. In the activistic, dynamic society
envisioned in the extrapolative scenario, none of these 14
major world problem areas is likely to be ignored and the
following ones are likely to be emphasized.

- World population growth and aging populations;

- Food: agricultural production and distribution;

- Foreign affairs and military security;

- Techno-economic security and viability;

- Energy; and

- Health and biosciences.

Likewise, under an extrapolative scenario, all of the ten representative emerging technological developments discussed in the report, and many more, are likely to receive increasing attention and programmatic support from both industry and Government. The ten technological developments discussed are:

- Birth control;

- Food: aquaculture;

- Health: combatting future cancers;

- Biosciences and bioethics: DNA (deoxy-ribonucleic acid);

- Microelectronics: computers and tele-communications;

- Transportation: short-hop (STOL) airliners;

- Technology-abetted political participatory systems;

- Energy: oil shale;

- Energy: nuclear fusion powers; and

- Space colonization.

Finally, under an extrapolative scenario, two organization policy alternatives are likely to receive increasing attention at the Federal policymaking level. These are:

- Further development of the Federal science and technology policy and management structures; and

- Further development, and perhaps institutionalization, of the Nation's analytical foresight capabilities and of effective linkages between those capabilities and technology policymakers.

To investigate these factors comprehensively, thoroughly, and continuously probably would require the institutionalization of the science and technology outlook at the Federal policymaking level.

Selected Findings

This section summarizes the RIAS findings about what is known, unknown, and hypothesized about industrial innovation and its role in the economy. It is presented in the form of a series of generalizations, observations, and hypotheses, omitting many important qualifications and detailed citations to the documentation. The intent of the section is to give the reader a feel for the nature of the innovation process, the mutual relationship between that process and broader socio-economic affairs, and some of the policy, institutional, procedural, and resource options available to alter existing situations or to create new capabilities for achieving desired goals. For more complete understanding the reader should refer to the full RIAS report. While all these statements have a reasonable probability of being correct, policy or programmatic action should not be taken without further analysis and verification.

With the above cautions and caveats, the RIAS team offers the following condensed list of generalizations, observations, and hypotheses organized into six categories:

(1) General statements about research and innovation;

(2) Operation of the (internal) innovation system;

(3) Innovation as a contributor and change agent in social, economic, and political affairs;

(4) Needs of the innovation system for resources, guidance, and stability;

(5) Institutional structure, codes, and procedures; and

(6) Information and research needs.

1. General Statements about Research and Innovation

- The United States is perceived as (and may actually be) losing global leadership and momentum as a technological innovator, which may contribute to falling national productivity growth and declining trade balances.

- There is a general loss of elan in the innovation communities of the United States.

- There is no coherent national science, technology,

or innovation policy and even the enunciation of
one by statute (which many feel necessary) would
be insufficient for major progress (the National
Science and Technology Policy, Organization, and
Priorities Act of 1976 (P.L. 94-282) is not adequate for this purpose).

- There is a lack of consensus on just what the
  "innovation problem" is.

- Each aspect of the so-called "innovation problem"
  must be dealt with in the light of other aspects,
  since we are dealing with a systemic situation.

- No single factor (e.g., executive or legislative
  Government, industry, university, consumer or
  labor group, financial institution) can do much
  alone to enhance innovation, but almost any one
  of them can retard it.

- Policy, institutional, and competing interests
  factors dominate both the problems and the solutions.

- Policy changes made now will have their major
  impact on society and the economy from 5 to 15
  years into the future.

- For constructive change to occur in operation of
  the innovation system, the following are needed
  simultaneously: widely accepted goals; understanding of the present situation, policies, and
  strategies for action; means for action (institutions, procedures, and resources), motivation,
  and time.

2. Operation of the (Internal) Innovation System

- Research and development are key elements in the
  innovation process and, in fact, often constitute
  a small proportion of total innovation costs.

- The innovation process operates in a far more
  holistic and interdependent fashion than is perceived by persons in the separate regulatory,
  policy and program-oriented institutions.

- The innovation process, like the proverbial ele-
  phant, is perceived differently by various groups.

For example, those listed in the left column tend to view innovation as listed in the right column:

Scientists and engineers	New technological concepts and capabilities
Business	Markets and profitability
Government	Utilization; solved problems
Economist (traditional)	Externalities, "residual"
Public interest groups	Possible negative social and environmental impacts

- The process of innovation may be viewed as a process of uncertainty reduction. The types of uncertainty include general business uncertainty, technical uncertainty, market uncertainty, and policy (including regulatory) uncertainty. High uncertainty impels firms to concentrate on innovations that are short term and incremental in nature.

- The underlying science and technology base is related to the pace of innovation in various industries. Innovation may be easier in industries with better-developed science bases and with technologies to which current "technological trajectories" are readily applicable.

3. Innovation as a Contributor and Change Agent in Social, Economic, and Political Affairs

- Industrial innovation may be regarded as a form of economic investment. There is widespread agreement that industrial R&D and innovation are important contributors to national economic growth and productivity.

- Through its effects on economic growth and productivity, and its contribution to new products and services, industrial innovation has a large impact on the quality of life.

- Due to uncertainty of innovation and problems in appropriating the benefits of innovative activity, there may be a tendency for industry to under-invest in innovation.

- Recent declines in the rate of economic growth, productivity growth, and trade balances cannot be tied conclusively to negative trends in industrial R&D and innovation. It is widely believed, however, that stimulation of innovation may be helpful in resolving today's economic problems.

- The most likely scenario for the future decade is an extrapolated (more of the same) pattern of scientific and technological innovation with major attention focusing on matters of population, aging, food production and distribution, defense, energy, health, and biosciences.

4. Needs of the Innovation System for Resources, Guidance, and Stability

- Innovation and economic growth are symbiotic processes, each contributing to the health of the other.

- Industrial innovation, like other investments, responds to economic forces, such as demand and costs. It is also affected by technical and institutional factors, which guide the direction of innovation and determine how rapidly innovation can respond to economic signals.

- Although it is commonly claimed that marked demand "pulls" innovation more effectively than technology "pushes" it, this represents an oversimplification. A market demand cannot be met if the technology to satisfy it does not exist or cannot be created. The key to successful innovation is the ability to match the demand with the technology.

- Generally poor economic conditions are a direct factor in the current malaise in innovation.

- To some extent, research and innovation activities require a stable environment with at least some cushioning from short-term fluctuations in policy, procedure, and resource allocation; some element of risk, however, is necessary to overcome inertia and stimulate innovation.

- The research and innovation "system" is significantly affected by Government policies, although these effects are often unintended.

- The turnover rates of capital, scientific and engineering project cycles, and institutional (public or private) performance standards time scales are out of synchronization and frequently operate negatively on each other.

- The reward structure for scientists and engineers involving peer recognition and professional status for technological innovativeness often conflicts with the entrepreneur's or investor's needs for near-term financial returns.

- The costs of innovation vary from industry to industry depending on the nature of the products produced, the capital-intensity and flexibility of production processes, and costs of innovation relative to the assets of individual firms. In some industries only the largest firms can afford to introduce innovations. Further, costs of different stages of the innovation process vary and thus R&D may account for a large part of innovation costs in some industries and not in others.

- The research and innovation climate is characterized by: constraints (Government regulations, capital shortages, reduced rewards for risk-taking), instability (general economic and political affairs, the unknowns of research, development and innovation per se), and misunderstanding (of the innovation process, of the relationship between innovation and the overall socioeconomic system, and among the multi-disciplined policy analysts and multi-oriented policymakers).

5. Institutional Structure, Codes, and Procedures

- Instability in Government policymaking, regulation, and political positions retards the private sector from making initiatives, particularly for the long term.

- The degree, rationality, and consistency -- i.e., the role -- of Government involvement in private research and innovation activities is a key issue.

- Adversarial relationships between the public and private sectors induced by our legal system (particularly between business and the Federal Government) hinders communication, understanding, and cooperation in devising policy and action alternatives; mutual trust and respect are missing among the parties.

- Institutional barriers and gaps among industries, within the executive branch, within Congress, and particularly among all the sets of institutions inhibit even partial solutions and new initiatives for innovation.

- Hardened institutional roles, bureaucratic inertia, and professional Balkanism hinders policy analysis and option finding.

- A growing tendency to search for someone or some institution to "blame" retards technological, marketing, and financial risk-taking but enhances issuance of regulations and protective control procedures.

6. Information and Research Needs

- More knowledge exists about the innovation process and its role in the economy than policymakers are aware of or use.

- At the same time serious deficiencies exist in the evidence and understanding underlying possible policy initiatives.

- The link between policies and national institutions at the level of project and firm (and even industry) activities at the micro level are poorly understood.

- Research findings are contradictory on the nature of relationships between technological innovation and firm size, industrial concentration, market entry, diversification, and spin-offs.

- The innovation process differs from sector-to-sector, industry-to-industry, and firm-to-firm. Analyses should be undertaken at these levels. It may be wise to consider industry-specific innovation policies or to allow sufficient flexibility to tailor programs to the needs of particular industries.

Implications for the Congress

The mission of the RIAS study was (1) to provide the Congress with a synthesis of what is known about the innovation process and its role in socioeconomic affairs; (2) to present this in the light of current actions, debate, needs, and opportunities; and (3) to offer some perspective on long-range outlooks for the future. Major analytical tasks remain for subsequent efforts. Nevertheless, the process of compiling this synthesis has revealed several major findings and options of immediate interest to the Congress.

The innovation "problem" has become a hot topic in the last couple of years, but it is not clear exactly what the problem is. A number of interrelated problems are frequently mentioned: a malaise, loss of elan, lack of incentive to take risks, etc. It is likely that there is actually a large, complex set of problems (and opportunities). It is difficult to find hard evidence on any of the problems. In addition to research findings, conclusions are often based on experience and anecdotal evidence. Although they present a mixed picture, trends in a number of indirect indicators of industrial innovation are consistent with warnings of a leveling-off of patenting and innovation in the United States and an increase in other countries - notably Japan and West Germany.

A major thesis of the RIAS has been that industrial innovation needs to be viewed "holistically" -- that is, as the sum of its interrelated parts and as a subsystem with the environment of economic, social, and political factors. The economic system guides and supplies resources to the R&D/innovation system (R&D/I), while the R&D/I system in turn contributes to economic growth and productivity. A variety of Government policies affect R&D/I including: patent policy, R&D funding, taxes, regulation, antitrust policy, and procurement. Government policies affect industrial innovation in complex ways that are not well understood.

A corollary to the finding that there is no single innovation problem is the finding that there is no "easy fix." We identified about 50 major studies and more than 500 policy recommendations. The recommendations seem to fall into the following categories:

- increase the flow of capital for innovative activities;

- improve the effectiveness of the patent system;

enhance the ability of United States firms to meet
international competition;

- improve the "technology base" that underlies many
firms or several industrial sectors;

- face the issue of Government incentives for individual (or small groups) of potentially healthy
firms in parallel with concern for weakened ones;

- actively and cooperatively search out regulatory
and other barriers to productive and innovative
endeavors and, at least on a trial basis, remove
them; and

- permit (experimentally at first) cooperative actions
among firms where the public interest is served in
parallel with the private interests of the firms.

One of the key findings of the RIAS is that policy for industrial R&D and innovation needs to be set with an understanding
of the R&D/I system and the manner in which Government policies
affect that system. This in turn leads to the need for a
"strategy" for innovation policy. The point is the need to
select a few key, mutually agreed upon, priority areas for
action and to use them to change the present trends and relationships. To be effective, each action must be taken in
the light of its impact on all the other actions.

Another key conclusion of the RIAS report was that there
is a need for parallel action to stimulate innovation at the
macro level of Government policy and strategy and at the micro
level of individual industries and firms. Action at either
level alone probably will be ineffective. Any improvements
in innovation and economic affairs must include efforts to
improve communications between the public and private sectors
and to develop means for cooperative analysis, planning, and
action. The situation is sufficiently polarized that no
sector or single institution can be effective through unilat-
eral action. Restoring mutual trust and respect; developing
cooperative analytical, policymaking, and assessment mechanisms; and creating a stable innovation environment are ad-
mittedly vague and difficult objectives. Nevertheless, this
study reveals no other more important or consistent underlying
factors.

The convergence of major innovation reports and discussions in the last couple of years has created a climate de-
manding Government action. The focus of that demand -- or
opportunity -- is now the Congress. The Congress seems to be

the principal party to take the initiative and to provide
leadership in such basic human and institutional matters. Of
course, the Congress cannot legislate trust, cooperation, or
stability. It can, however, take a number of actions that
can move toward an improved climate for industrial innovation.

It appears that only two options are really "closed" to
the Congress -- to do nothing and to launch another major
long-term study. This leaves a wide variety of congressional
options that are open. Broadly, they fall into five categories that can be approached in parallel:

(1) concerted action on the menu of bills related to
innovation that have already been introduced;

(2) reacting to proposals from the executive branch --
especially the President's Message on Industrial Innovation;

(3) reviewing the several major innovation studies
(including this one) to identify actionable items and to develop a broader perspective on innovation;

(4) initiating steps to improve and stabilize the
general climate, both for technological innovation in the
private sector and for policy and program innovation involving the public and private sectors; and

(5) initiating broad foresight and assessment activities
for systematic and continuing action coupled with a few selected analytical efforts to answer key questions identified
in the other four areas.

Within each of the five categories, the range of options for
congressional action is considerable. Depending upon one's
definition of "related to innovation," there are between 20
and 200 such bills before the 96th Congress. The President's
Message on Industrial Innovation made 31 specific recommendations, at least half of which invite some form of congressional involvement. Based on early reactions (and excepting
patents), congressional interest seems to be primarily in the
President's Message. A detailed analysis of the more than
500 policy recommendations from existing innovation studies
could also generate a variety of additonal options for congressional consideration.

It is sometimes suggested that the Government identify
specific areas of technology for priority attention. Examples
of the concerted efforts of the Japanese in video recording
and now in robotics are often cited. Similarly the United
Kingdom is focusing on selected aspects of information

technology, and Sweden on forest products, among others.
There is no consensus nor satisfactory analytical argument as
to which set of technologies the United States should opti-
mally pursue. Moreover, there are strong feelings and persuasive arguments for leaving the decisions to the market
forces, domestic and global. There appears to be neither a
need nor the ability for the Congress to select a few "lead"
technology areas for the United States to pursue on a priority competitive basis in world markets. Congress, along
with other governmental institutions, can and will continue
to establish priorities for innovation activity for meeting
public needs and desires. In addition, the Congress has the
opportunity, and what many feel to be a growing responsibility,
to act to improve the general climate for innovation as a
contribution to improved economic well-being.

It may be that the nation is approaching the point where
some action, even if imperfect, may yield far better returns
than continuing the present pattern of waiting until sure, or
reacting suboptimally to a variety of crises or special interest demands. The pressures for action are mounting. The
knowledge to take a variety of experimental and some permanent actions exists. The rewards in increased social and
economic well-being would seem to exceed the risks of partial
failures and greatly exceed those of maintaining the status
quo. The Congress may be the only institution which can provide the initiative, leadership, and means for cooperatively
"solving the innovation problem" and revitalizing this key
aspect of our economy. It is hoped that the JEC/SSEC/RIAS
synthesis of what we know, where we are, and perspectives on
current and future options will assist the Congress in that
role.

6. The International Dimensions of Innovation

The Organization for Economic Cooperation and Development (OECD) is the international economic organization which includes the 24 governments of the major industrial democracies and market economies. It promotes policies of growth, employment, rising living standards, and related world development based on the maintenance of a liberal multilateral trading system.

The following views on innovation policy, while personal, reflect the results of studies in the OECD on

- positive adjustment policies to encourage industrial growth through research and development and innovation;

- the role of science and technology in relation to present and future conditions of growth in the OECD countries;

- the effects on national science and technology policies of transfers of technology among OECD countries and between them and the newly industrializing countries and the non-market economics of Eastern Europe;

- factors that influence innovation in medium and small enterprises;

- the impacts of multinational enterprises on national scientific and technological capacity.

Given that technological change is the key element in economic growth, OECD countries attach considerable importance to technological innovation. This is underlined by a number of new policy measures recently announced, and by the

scale and priority of the national studies which are currently underway. These include President Carter's message to the
Congress on innovation measures and corresponding comprehensive approaches or studies by the Netherlands, Sweden, the
U.K. and Canada, Japan, among others.

Although these concerns sometimes find expression in
assumptions that the rate of technological innovation has
slowed down, such slowdown in innovation has not been demonstrated, nor is there an accepted methodology for the
measurement of innovation. Nor is it necessary to rest the
case for accelerating the rate of innovation on the grounds
of a slowdown in innovation. The need to accelerate the rate
of technological innovation rests on other arguments, among
others:

1. the present and prospective slow economic growth
    and high unemployment;

2. the need to improve balance of payments conditions
    through stepped-up exports. For OECD countries,
    these are largely characterized by product unique-
    ness rather than price competition -- a reflection
    of innovative content;

3. the need to promote innovation in new products to
    provide increased employment opportunities generally
    and achieve the needed increases in productivity;

4. the increasing role in international trade played by
    newly industrializing countries who are building
    their future export positions on the combination of
    technological sophistication, low labour costs and,
    in some cases, preferred access to raw materials;

5. the international flow of technology among OECD
    countries, North-South and West-East, requires a
    continuous flow of knowledge-intensive innovation
    in OECD countries to stay at the front end of the
    product cycle according to national comparative
    advantage.

Given the need to accelerate the rate of desirable technological innovation, many OECD governments are devising
policies and programs to strengthen and supplement the market
forces which normally provide a powerful stimulus to research
and development and innovation. In devising such policies,
it is important to differentiate invention from innovation,
while recognizing that the innovation process reaches from
long-range research to the ultimate user. Invention and the

research and development on which it is often based are inputs to innovation. On the other hand, innovation is the introduction into manufacturing or the market of something new. Since innovation is on the output side, the forces that influence it are usually external to the enterprise -- the economic climate and other non-technological factors influencing risk. While R&D and invention are inevitably influenced by the external climate, the factors exerting more powerful influence on research and development are largely internal to the science and technology sphere. Failure to distinguish adequately between these two different but related policy spheres has often resulted in lack of balance in policy formulation with preponderant emphasis given to the strengthening of R&D and inadequate attention given to the external climate necessary to encourage innovation.

The single most important and often essential precondition for innovation is ability and willingness to invest. At present, a more worrisome aspect is the general lack of propensity to invest. A reasonably strong rise in aggregate demand and the prospect of sustained growth would, according to macroeconomic demand management theory, provide a sufficient stimulus for investment in innovation. But there are strict inflationary limits on macroeconomic demand management. Even with a sustainable increase in demand, structural and other factors will limit investment in innovation. Since investment decisions involve consideration of both risk and return, policies that can reduce the level of risk and can improve the rate of return on investment will be of obvious benefit. Risk and return, particularly in the short term, are critical to investment decisions since companies normally aim to maximize the discounted present value of future net cash flows.

However, it should be noted that there can be a high rate of innovation in a specific sector where the demand is high, even though the aggregate demand is low. This characterizes particularly the electronics sector -- microprocessors and computer-based information and communications systems. In time these innovations will diffuse throughout the manufacturing and service sectors with job creation and job displacement effects of as yet little understood proportions. With their short-term labour-shedding bias, it is ever more important to assure increased employment opportunities through innovations offering new products and services.

Experience has shown that incentives to invest in innovation can most effectively be provided through general economic and fiscal policies rather than by way of selective measures. Policies to encourage investment in new plant and equipment have long been used. There is, as yet, little attention to experience with economic, fiscal, tax, and related policies specifically tailored to the promotion of innovation, although this is clearly an area requiring examination. A major advantage of such general measures to stimulate innovation is that they can reach a multitude of decision centers and encourage innovative management and investment throughout society, adapted to local needs and the timing of individuals and groups.

I have made reference to structural factors influencing the extent and direction of innovation -- factors that can impede or distort impulse to innovate that can be expected to accompany an increase in demand. Such structural factors are listed below, not necessarily in order of priority. Each is of such importance and complexity that they warrant more extensive treatment than can be given in this brief text.

Structural Factors in Innovation

1. Regulatory Policies

It is well known that regulatory policies can both stimulate and inhibit innovation. Regulation in the environmental, health, and safety areas is required in the public interest, and the costs associated with meeting necessary regulations must be accepted as costs internal to the development of a publicly acceptable product. They should not be regarded as externalities or a diversion to "defensive" research and development. Nonetheless, it is important that the regulatory costs and delays in associated administrative procedures be kept to a minimum consistent with public safety. This requires science-based regulation and the avoidance of unnecessary regulation and reporting requirements as well as better coordination, clarity, and continuity in government regulations. The concept of science-based regulation is extremely important since relatively small reductions in regulatory requirements consistent with improved understanding of health and environmental impacts could make considerable differences in product or process costs. Equally important to risk-taking is that regulations be reasonably stable and predictable so that there is general confidence in the regulation process.

Another aspect of regulation concerns regulated or nationalized industries in many countries, such as communications and transportation. Here, it is necessary to pay special attention to assuring the existence of sufficient incentives to promote innovation. There are two principal dimensions. First, regulation can stifle competition neces-

sary to stimulate innovation. Second, nationalized industries can also be non-innovative if they are unnecessarily
subsidized in terms of what is required to meet the disci-
plines of the marketplace.

2. Protectionism

It is well known that industries that are protected from
international competition by tariffs, quotas, or other trade
measures have a relatively weak performance in innovation.
In OECD countries, this has characterized sectors such as
textiles and footwear.

Progressive reduction of protective barriers and expo-
sure of companies to competitive market forces can be a powerful stimulus to innovation and the establisnment of healthy
and viable enterprises, even within sectors that are generally characterized as declining in favor of countries with
lower labor costs.

3. Technology Transfer Policies

Technological innovation is best fostered in a liberal,
open trading system that permits technology to flow internationally and be incorporated in manufacturing, in products,
and in services according to relative advantage in factors of
production. Policies that unduly interfere with this flow by
restricting technology export or unduly subsidizing local investment by foreign technology-intensive companies, or which
discriminate in favor of local technological enterprises and
services, could cause undesirable distortions of the innovation process. While there is, as yet, no major cause for
concern on this score, a prolonged period of slow growth and
high unemployment could provide pressure in this direction.
Some recent policy suggestions in the European Commission and
elsewhere in Europe, as well as in the U.S., give cause for
disquiet on this score.

4. Manpower and Social Policies

Innovation to increase productivity is essential to promote economic growth. While the immediate effect of such
innovation may be labor displacing, the second order effects
will be to generate more new jobs through the resultant
economic growth. Properly designed and implemented manpower
and social policies are needed to facilitate the introduction
of labor-saving innovation and to avoid negative social feedback that could slow or stop economically and socially desirable innovation. Such policies must recognize that the skill
levels required for new jobs are generally higher or differ-

ent from those displaced. Policies are also needed which encourage mobility throughout society - horizontally as well as
vertically.

5. Patent Policy

The original basic thrust of patent law to promote progress in the arts and useful sciences and the disclosure of
resultant inventions has shifted. With the emergence of
large industrial R&D laboratories, patent protection has become an increasingly important means to reduce risk and promote investment in innovation. For small- and medium-sized
enterprises and individual inventors, patents are especially
important to investment and market entry. Thus, the maintenance of a strong patent system and the issuance of patents
of high validity are important elements of national innovation policy.

6. Small- and Medium-sized Firms

Small- and medium-sized firms are especially important
as a source of innovation and new employment, as well as to
the maintenance of competition in innovation. In a period of
slow growth and a lack of propensity for investment and risk-
taking, the small- and medium-sized enterprises are endan-
gered. This danger is magnified by the necessity and cost of
meeting increasingly stringent and numerous government regulations. The small science-based company looks more and more
to the large company for longer-range research and development -- where there is a disturbing tendency to neglect potentially important innovations in favor of alternative investments along established product lines. The very nature
of technological innovation is changing with the increased
sophistication and large systems characteristics that have
increased the cost of entry to the disadvantage of medium and
small enterprises. Further, large companies are acquiring
smaller growth companies, thereby enhancing oligopolistic
tendencies. Thus, special policies are needed to stimulate
innovation in SMEs to assure access to seed money for early
design, risk capital, and necessary technical information.

7. Innovation in the Service Sector

There are certain areas where markets are unlikely to
adequately reflect and anticipate future economic and social
needs. This applies, for example, to research and development and innovation in producing and saving energy, to improvements in environmental quality, health care, urban infrastructure, etc. Specially designed innovation policies
are required to set performance standards and to aggregate

markets so as to encourage competitive R&D and innovation in
satisfying the latent demands of the public for improved services at reasonable cost.

8. Diffusion of Innovation

Due to the lack of diffusion of innovations within a
given sector, there is a large difference in performance between the best and the average firms in most industrial
sectors. It is also established that many important innovations in a sector, say textiles, originate outside of the
sector in question, in this case, the chemical industry. Few
studies exist on diffusion; what we do know is that it is
cumulative and that the small firm has a particularly important role to play in its promotion. Thus, policies are
needed to encourage the diffusion of existing innovations
within and between sectors. This is a very important but
largely neglected area of innovation policy.

9. Public Acceptance

There are important technologies whose introduction into
society is slowed by lack of public acceptance. The classic
case is that of nuclear power, but there are other cases.
Where lack of acceptance is due to real uncertainties concerning safety, environmental or social impacts, the slowdown of
innovation is socially desirable. On the other hand, public
reluctance may be due to unreasonable fear; the more complex
a technology, the fewer people understand it. There are now
requirements for public understanding and participation in
decision-making generated by certain major technological
advances that call for special attention in government public
information programs and administrative procedures to facilitate the introduction of socially desirable innovations.

10. Direction of Technological Change

The direction of technological change, particularly
when it causes imbalances in societal efforts, can give rise
to structural problems that could operate against further
innovation. We have seen this in the concern of the '60s
about the effects of automation on employment. Similar concerns are now evident in the rapid development of computer-
based information systems and microelectronic developments,
particularly the large-scale introduction of microprocessors
in industry. There is, at present, a labor-saving bias in
innovation that, while necessary to encourage from the stand-
point of productivity, must be balanced by measures to foster
a broad range of product innovation and to deal satisfacto-
rily with the labour displacement consequences.

11. Scientific and Technical Inputs

The foregoing structural factors bear on the output side
of innovation. There can, of course, be structural impediments on the input side; impediments to the generation and
flow of scientific and technological developments to provide
the basis for innovation. There have been many papers
written on the subject of market pull vs. technology push as
forces for innovation. Of course, both are important. Without a market, innovation is bound to fail. But the market
may not exist until the technological possibility is known.
Or the market may be weak due to the general economic climate,
out of phase with the 8-10 year lead time for many technological innovations. This situation has prompted governments
to intervene to supplement market forces or to facilitate the
long-range development of major technologies where appropriate in the public interest, e.g., energy technologies. Without such intervention, the innovation process may not take
place at all or in time to serve the public need. The justi-
fication for government intervention increases with the
"distance" from product development, being greatest in basic
and exploratory research where no single company may be able
to capture sufficient benefit from introducing the innovation
to justify the decision to invest.

Selective Intervention

I now turn to the question of the way in which governments intervene to stimulate innovation. It should be emphasized that under normal conditions there is usually a pre-
sumption against selective action to assist particular
sectors or companies in strengthening their innovation capabilities and in favor of more general measures to improve the
economic climate conducive to investment and risk-taking. On
the basis of past experience, general measures have exerted
much more effort on innovative performance than direct assistance by governments with the possible exception of energy,
where government intervention on a broad front was dictated
by public necessity.

Yet many OECD governments have chosen not to follow a
policy of merely letting good firms succeed and less successful firms fail. Accordingly, it is necessary to consider the
manner rather than the desirability of more direct forms of
government intervention.

1. Enterprise Level

In most industrialized countries, private firms are ready
and able to finance those innovations which they believe will

be profitable. There is real danger that when subsidies are requested, they will be for less satisfactory projects. A series of innovation studies have demonstrated that the most common causes of failure are not those associated with lack of finance for development but those related to a poor understanding of the market. The availability of generous R&D and innovation subsidies seems far more likely to reinforce this pattern of behavior. This would suggest that more efficient and socially desirable forms of policies for the stimulation of innovation might emerge from procurement policies rather than from R&D and innovation subsidies.

2. Sectoral Level

The arguments for government finance for commercial technology ventures are strongest where the market is particularly weak, such as in highly fragmented industries, and where there is strong public need. Subject to the legitimate constraints of antitrust policy, there could be a case for encouraging cooperation among companies in demonstrating the economic feasibility of technological innovations that could benefit a sector as a whole.

3. "Picking the Winners"

In some member countries, industrial structures are not yet sufficiently developed to sustain the costs of an ambitious "high technology" venture of particular national importance; in others, the predominance of small and medium firms may render the market vulnerable to a foreign technological oligopoly. In such cases, state intervention may be the only way of building the technological infrastructure to support development in a particular sector. However, if the infrastructure is sound, and if the market is working properly, projects to develop technological products for the market place which cannot command commercial support risk being only "second-best" products. Governments which engage in them may therefore be inviting failure unless, through use of their own special powers, they are in a position to ensure success. If more than one government chooses to back the same "winner," the result can be an international contest whose repercussions can go well beyond the bounds of science and technology policy, engaging difficult and potentially disruptive political issues. Such contests may lead to over-capacity, or accusations of unfair competition.

The recently agreed GATT Code on Subsidies and Countervailing Duties does not restrict the right to use subsidies to achieve important policy objectives including the objective to encourage R&D programs, especially in high technology

industries. However, it cautions signatories to avoid causing injury or prejudice when drawing up policies and practices in this field. In view of the increasing financial
assistance by governments in such areas as the microelectronics industry, this section of the Subsidy Code may well be
pressed in time for further amplification and clarification
to avoid differing national interpretations.

Scientific and Technological
Inputs to Innovation

Since a necessary condition for innovation is the availability of the requisite scientific and technical (S&T)
potential, a prime function of effective scientific and technological policies is to ensure that the requisite S&T potential is available. This implies maintaining healthy national
research systems, facilitating as much as possible the flow
of technical information between the public and the private
sector, ensuring conditions which provide adequate supplies
of qualified manpower, and securing access to technology from
abroad.

Beyond assuring a healthy scientific and technical infrastructure, government policies are directed to the promotion
of basic and applied research in the universities and non-
profit institutions to provide a science and technology knowledge base as well as the advanced training of scientists and
engineers.

Government support or encouragement of R&D in industry
for commercial purposes is not as well justified and accepted.
This has been particularly true for the United States although there has been commercial spin-off from government
support of military and space research and development in
private industry. On the other hand, government support or
encouragement of R&D in industry for commercial purposes is
felt to be necessary in many European countries, Canada, and
Japan.

There are hazards in government encouragement of research
and development in industry for commercial purposes as con-
trasted with the situation where the government is the customer. In the latter case, the performance characteristics
are generally known and the market is certain. The absence
of these characteristics in commercial-market-oriented R&D
support by government makes the venture much more speculative
and less certain of success.

In any event, it is important not to over-centralize
the R&D decision-making process. Multiple decision centers

remain a necessary and desirable "hedge" against technological
risk factors.

Government Policy and Structures
for Innovation

It follows from the preceding remarks that a wide range
of government actions - many undertaken for quite different
purposes - have direct and indirect effects upon innovation.
Though perhaps less clearly perceived, these indirect effects
can be just as important as the direct effects of measures
taken for the specific purposes of promoting innovation.
They can create incentives for innovation or barriers to it.
Thus, they must be identified, measured, and evaluated as an
integral part of government innovation policy.

1. Technology Policy

Whereas policies for science have carved a firm niche in
the hierarchy of government structure and policy-making, the
same cannot be said for the area of technology policy which
extends throughout the mission-oriented structure of government. Yet a strong case can be made for central oversight
of technology policy at the highest level of government to
assure that the generation and deployment of technology for
public purposes provides general economic and social well-
being. Innovation policy affords a graphic example of the
reason and rationale for such a central technology policy
function -- recognized as distinct from science policy in its
conception, application, and administration. Yet it has yet
to make its mark in macroeconomic thinking with which it must
be intimately bonded.

2. Integration of S&T and Economic and Social Policies

This underlines one central theme that has emerged in a
number of studies by the OECD Committee for Scientific and
Technological Policy: this is the urgent need for better
integration of the scientific and technical aspects of public
policy and the social and economic aspects. The need is now
greater than ever and has been heightened by the drive to
accelerate innovation and technological change. The scientific and technical aspects of policy must be guided by
objectives usually formulated in social and economic language
and coordinated with other dimensions of policy. The more
traditional social and economic policies must be employed in
a way that recognizes scientific and technical opportunities
and constraints and exploits and encourages scientific and
technical possibilities.

International Cooperation in
Promoting Innovation

One country's innovations can affect the adjustment needs and policies of others. Both the need for adjustment and the possibilities for achieving it are strongly influenced by the underlying pace and directions of technological change, by its rates of diffusion within different national economics, and by the ease with which technology is transferred from one country to another. These are factors which are international in scope. Action by governments acting singly can have -- and has had -- considerable effect on the pace and directions of technological change. But there are limits beyond which no government acting alone can control or moderate the international rhythms or the spread of technical innovation. The countries of the OECD have become partners in a world system of dynamic interdependence based on continued innovation and the unimpeded flow of technology within and across frontiers. Perhaps the most important medium-term priority is how to live in this dynamic situation, when the tools with which we have to work were mostly designed to cope with slowly changing opportunities and international relationships.

Role of the OECD

The reserve of possible governmental policy instruments to stimulate innovation is quite large. Some have been tried out extensively, others are under scrutiny (e.g. in social innovation), and there are unused ones, particularly in the class of indirect yet specific aid. The precise combination of policy instruments must necessarily vary both between industries and between countries, and it is important to recognize the limitations of government interventions as well as their advantages. To multiply forms of inefficient and bureaucratic interference would stifle rather than stimulate innovation. Nevertheless, all OECD countries have found it necessary to use a variety of policy instruments, and the exchange of experience on their effectiveness is a valuable feature of international cooperation.

In support of such exchanges of experience on national innovation policies and in extending them to the discussion of compatible policy approaches where international harmonization is desirable, the OECD is taking steps to extend its international statistics on R&D expenditures to the development of internationally comparable output indicators including those that measure performance in innovation. Transparency of national policies and measures for the stimulation of innovation is also an important objective of OECD cooperation. Each member country has much to learn and a little to teach.

In conclusion, although governments have shown increasing interest in policies for the stimulation of innovation and have been willing to introduce many new measures, they have devoted little attention to studying the relative efficiency of such policies or their impact. Comparative studies could throw much light on this question and possibly prevent some misallocation of resources. International cooperation in this area is in its infancy. Like most infants, it expends more effort in learning than in efficient performance. But it is an infant which needs to grow up quickly and whose education all countries must provide.

Discussion: U.S. Technological Policy Needs:
Some Basic Misconceptions

Development of constructive approaches to strengthening the international competitiveness of domestic industries requires revising five misleading conceptions of the fundamental problems to be dealt with.

First, contrary to common beliefs, the foreign challenges to domestic producers are no longer focussed primarily on displacing marginal industries characterized by relatively simple technologies and small-scale operations. Instead, they are being increasingly directed at the very industries whose superiority has seemed most secure because of their technological complexity, large-scale operations, and management sophistication. The targets have accordingly reached beyond textiles, shoes, and clothing to encompass automobiles, steel, machinery, and consumer electronics. Moreover, serious thrusts are also imminent in respect to such "high technology" domestic industries as large- and medium-size computers, semiconductors, aircraft, and pharmaceuticals. And such foreign development efforts are aimed not only at increasing dominance of their own home markets, but also at gaining, and then retaining, as large a share as possible of U.S. markets.

Widespread failure to recognize these bold aspiration levels of foreign producers helps to explain the repeated reactions of surprise when strong foreign competition has emerged in industry after industry which had unconcernedly observed the discomfiture of earlier victims. Serious analysis emphasizes that the problem to be confronted is not as simple as has been suggested by those who have proposed abandoning such "backward" industries in the expectation that offsetting gains are available by merely intensifying our commitments to "higher technology" sectors. On the contrary, it is necessary to recognize that the foreign challenge affects all industries including those developing and utilizing the "highest" technologies and that major technological advances are attainable which can yield substantial competitive advantages in virtually all industries. In addition, sound evaluation of the importance of seeking to strengthen the technological competitiveness of various domestic industries must take account not only of the sheer novelty or sophistication of prospective further technological advances, but also of their potential short-term and longer-term impacts on domestic employment levels, on their contributions to reinforcing the competitiveness of the industries utilizing its products, and even on national security.

Second, the declining international competitiveness of an increasing array of domestic industries is not due only to the inevitably more rapid growth of hitherto underdeveloped nations on the basis of increased education, resource development, and access to capital. Rather, it is due in greater measure to the expansion of efforts to improve the competitiveness of already advanced industrial economies, especially through the greater encouragement and support of their governments. Another significant cause has been the declining commitments by domestic producers to strengthening the technological foundations of their competitiveness. In short, the problems result not only from the inevitable and patently unrestrainable surge of industrialization in the hitherto less developed parts of the world, but to a larger degree, from the contrast between the intensification of major technological improvement efforts in other advanced economies and the relative diminution of such commitments by many domestic producers.

Third, such decreasing commitments are not due to managerial ignorance, indifference, incompetence, or sloth. They are traceable to analyses suggesting that a variety of alternative allocations of resources offer more attractive financial rewards than programs to improve technology and productivity. Hence, there is good reason to doubt the adequacy of frequent proposals to ensure an increase in technological improvement efforts by simply increasing depreciation allowances and decreasing tax rates. It is reasonable, of course, to claim that increased profits and depreciation might encourage greater allocations to technological improvement undertakings, although such innovational efforts are also not infrequently stimulated by decreasing profits and increasing threats to survival of the firm. But, the belief that increases in cash flow _would_ generate comparable increases in needed innovational efforts is open to more serious question, unless the financial incentives for doing so were substantially increased relative to those offered by alternative allocations of any increases in available resources.

Fourth, technological competitiveness cannot be restored quickly. Even if major improvements could be effected without delay in the strength of incentives to enhance the technological capabilities of major domestic industries relative to concomitant deterrents, it would take at least 5-10 years for our already lagging industries even to catch up with leading foreign competitors -- and surpassing the latter through development and utilization of major advances in technology may well take much longer.

Finally, it is unrealistic to concentrate solely on the constructive potentials of advances in technology and productivity -- including increased economic growth, higher standards of living, more effective adaptation to resource stringencies, and the easing of inflationary pressures. Underlying such eventual aggregate benefits, one must recognize an array of more specific and more immediate effects, some of which tend to be less universally welcome. Hence, efforts to achieve more effective advances in the technological capabilities of domestic industries must be based on a realization that such gains have been, and may well continue to be, subject to significant resistances as well as stimuli.

In addition, I would like to offer the judgment that no amount of peripheral improvisations around current macro-economic or micro-economic policies is likely to prevent the continuing long-run decline in the international competitiveness of major domestic industries. A reversal of this trend requires revitalization of the technological capabilities of such industries. And this can be accomplished only if it becomes a primary target of new micro- and macro-economic policies instead of remaining a happenstance by-product of policies dominated by other aims.

Discussion

This is the second time in the last couple of years
that I have been in the position of following Bela Gold on
the program. It has its advantages and disadvantages. The
advantage is he wakes everybody up. The disadvantage is that
anything else is an anticlimax. I will follow his lead and
cover five points in ten minutes (although I carefully
counted his and he mentioned six, but that was his privilege).
I will touch on several matters that were either not fully
covered this morning or to which I would like to draw particular attention.

One notable point that was made -- Dave Beckler made
this point -- is that innovation is a matter of concern in
all developed countries, not just the United States. As we
carry on this great internal debate, we should be aware of
the fact that it's paralleled elsewhere. The relationship
of technology to economic growth has been documented in many
research studies. But governments are inclined to show interest in the subject only when they perceive problems that
voters care about, and their willingness to do something depends upon political tradeoffs. Actions may be called for
to promote innovation which are perceived by the government
as having penalties in some other sense. The statement that
Hilliard Roderick made in his comments from the audience --
that he's heard all this before, 10 or 20 years ago and why
didn't somebody do something then -- can be answered by saying that it's precisely for that reason.

The concern of the U.S. has been reflected in the Domestic Policy Review, which Dick Meserve discussed. The reaction to it has been mixed, at best. On one hand critics
praise it for having at least demonstrated interest on the
part of the present administration. The critics, however, go
on to say that with respect to the actual content of the Review, it doesn't address the most important issues. Most

significantly, as Dr. Meserve pointed out, it doesn't address
the question of tax policy, or, more broadly, economic and
tax policy. We are told to wait -- that will come later. In
the meantime, then, one has to withhold judgment.

People generally agree that economic and tax policy is
the number one issue, but regulation is perhaps second on the
list. The D.P.R. paid attention to that, but in generali-
ties, so one has to wait and see what actions will follow.
There were lesser matters, like patent policy, cooperative
research, procurement, and so forth, which are good but unlikely to have major impact.

The D.P.R. gave central attention to the development and
transfer of knowledge rather than to its use. This is attractive to government because it can be done merely by spending more money, which is what government does best. This
receives praise from some, because they expect to be recipi-
ents of this extra spending; it gets opposition from none, because it does some good even though it doesn't address the
central problem, which is the reduction of barriers. But the
generation and transfer of new knowledge, the central thrust
of the D.P.R., ducks the real issues.

To go on to a second point, and this may seem a little
strange from a representative of industry, it's fashionable
in meetings like this to talk only about things that result
from government policy. Much less attention has been given
to internal factors within companies and yet these are important. Perhaps it's hard to do research on this. There may
be some evidence, however, that top managers in many industries have a shorter term outlook, in part, because of market,
investor-related, and internal factors for the company, even
though the change may be ascribed solely to government policy.
The effect on the bottom line of a long-term outlook on investment in R&D and innovation may be negative in the short
term, even though it is strongly positive in the long term.
Management's willingness to make such investments thus depends on its evaluation of the importance of this year's
bottom line as compared with future bottom lines, and on
superficial things, like the words of security analysts,
stock market reaction to earnings reports, incentive compen-
sation schemes, etc. We don't understand well enough the
importance of these matters.

Third, and this point has been emphasized by other
speakers, the U.S. economic climate discourages savings and
investment. Neither individuals nor corporations are encouraged toward savings by government policies, which encourage consumption. By contrast, savings is encouraged in

Germany and Japan, two principal international trade competitors. Any slowing of innovation in the U.S. can be traced rather directly to our low savings rate.

Fourth, national attitudes on pride in work output and the quality of work have reflected a steady decline. When this is compared with Japan, one is struck by the superiority of worker attitudes and the quality of products from Japan. This is not just a criticism of U.S. labor force - it's a combination of national attitudes and those of both labor and management.

The fifth point relates to government attitudes toward the private sector. Professor Hambraeus discussed this, also. Government grows -- one intervention by government in the private sector leads to another, and an increasing proportion of our national resources is absorbed by government. It takes upon itself activities that formerly were carried on in the private sector. It subsidizes institutions in the private sector that exhibit an inability to compete, and at the same time attacks and seeks to reorganize those parts of the private sector that do compete successfully.

As a related matter, government seeks to support and direct R&D efforts for commercial markets. But the government simply cannot manage things well when it comes to commercial markets. It lacks a feedback mechanism, so it doesn't know what is succeeding and what is failing. It can tolerate inefficiency, because it can print money. The government is good at R&D support when it is the customer, as in procurement for the Defense Department, and its record there is in stark contrast to its record in energy, for example, where it is not the customer.

Finally, to emphasize a point made by Dave Beckler, there are conspicuous differences among industries - one shouldn't just lump them all together. Electronics is the outstanding example of an industry where there hasn't been a decrease in innovation. And the other extreme is a number of industries - basic metals, paper, and some others - where there hasn't been much innovation for a number of years. And, there are some industries where there appears to have been a change in the last few years - the chemical industry, for example. It's quite clear that you can't generalize, but you have to look at things on an industry sector basis.

I will conclude by re-enforcing the point that David Beckler made, which is that even though various industries are behaving differently, and we can't prove a case for a general decline in innovation, we cannot lose interest in our

rate of innovation and in policies that affect it. What we
really ought to be asking is whether industry can't do much
better than it's now doing, and on that I believe there
should be no disagreement.

Discussion

One of the disadvantages in speaking last is that some of the things you want to say have already been said. I hope to open up on a heretical note. Let me try it.

It is regarded as axiomatic in discussions of technology policy that we are doing very badly, and yet it's not totally obvious that this is so. By some criteria we are doing, at the very least, much better than many people believe.

First of all, the growth in American real per capita income was virtually as high through the decade of the 1970s as it had been in the 1960s. You're surprised at that, and I confess that I was too, but it does seem to be the case - just about as high.

Secondly, the economy has performed exceedingly well in the creation of new jobs. Indeed, part of our problem derives from having generated new jobs at such a rapid rate. We've introduced people of lower skill levels into the labor force very rapidly, and this is certainly one of the reasons for the stagnation of labor productivity growth. There are other reasons of course. One of these is certainly the low rates of capital formation in the U.S., as a result of which the rise in the capital-labor ratio in America has lagged considerably behind that of many other countries, particularly some of the countries which we regard as our most serious competitors in world markets.

Third, in absolute terms, total American R&D expenditures are still greater than those for all the other OECD countries combined. At least that was the case for the last set of numbers which I saw for 1976. What seems to be at issue is not so much a decline in R&D expenditures as the fact that the productivity of R&D activities, at least as that productivity is ordinarily measured by increments of

output, has declined. Let me suggest one reason for this
apparent decline. We are obviously manifesting greater
interest as a society in things other than economic growth.
We no longer pursue an expanded output of material goods with
the singlemindedness which we once did. Our very success in
attaining economic growth seems to be tinged with frustration
or, at the very least, disappointment. It isn't everything
we thought it was going to be -- it extracts significant and
unexpected costs. We are disenchanted. The reasons for that
are various, but I'd like to mention something which I think
is compounding it -- inflation. Inflation has very insidious
psychological effects. Everybody looks at this situation and
says, "Gee, whiz, with my present income, if prices hadn't
risen the way they have in the last 10 or 15 years, think of
how well off I would be," never considering the fact that his
or her present income is itself the result of all the inflationary pressures of the past decade or so.

        It is, I believe, a serious mistake to concentrate too
much upon the decline in R&D as the cause of the poor productivity performance of the economy over the past decade or
so. Of course, that relationship is an important one, but
it's essential to realize that causality goes both ways. The
willingness of industry to undertake R&D depends in fundamental ways upon their economic outlook, upon their assessment
of the future prospects for the economy in general and for
their own section in particular, upon the weight of the uncertainties that they have to contend with such as energy
prices and energy availability, upon how their long-term
plans are modified by inflationary expectations, etc. So, it
seems to me that if we want to link up R&D with economic performance, then we need to look with particular care at the
component of R&D which is privately financed and not focus
attention so exclusively, as many commentators do, upon
what's going on in the public sector, in public support of
R&D. Economic performance in the short run is more affected
by private R&D spending than by public R&D spending. And
this is a component which is strongly shaped by economic
expectations and by calculations of private profit return.
It's terribly important, it seems to me, that we take measures to bring our economic house in order because by doing
so it will have the effect of very considerably strengthening
private R&D spending. I am thinking of dealing more effectively than we have with inflation. I am thinking of, as
some speakers have just mentioned, the weak incentives which
now exist to investment and saving especially in our tax
structure and of the overall impact of environmental and
other regulations upon the incentive to invest in R&D. The
obvious failure, the disastrous failure to deal more effectively than we have been dealing with the energy problem.

I am very impressed with the fact that the most intense competition which we are confronting in international markets
comes from two countries where a much higher percentage of
all R&D activity is privately financed. I cannot help believe that that is a very significant fact. In both Japan
and Germany the proportion of R&D activity which is privately
financed is vastly greater than that of the U.S. It seems to
me that this has a great deal to do with the overall incentive system within which business functions in those countries, and I think it ought to provide some food for thought.